4-18-98
To Roger,
Thank you so much for your support!
Angela

OUT IN THE OPEN...

THE ABUSE OF INJURED POSTAL WORKERS

OUT IN THE OPEN...

THE ABUSE OF INJURED POSTAL WORKERS

Angela V. Greene

(202) 737-0989
(202) 737-0984 FAX

ISBN 0-9659070-0-7 (Soft Cover)

Copyright © 1996 by Angela V. Greene

All rights reserved. Neither this book, nor any part thereof, may be reproduced in any form without the written permission of BRAVA PUBLICATIONS, a division of BRAVA PRODUCTIONS, INC.

Library of Congress Catalog Card Number: 97-094208

PRINTED IN THE UNITED STATES OF AMERICA

"Do Unto Others As You Would Have Others Do Unto You."

FOR MY MOM AND MY DAD

They loved all people

ACKNOWLEDGEMENTS

Most of all, I acknowledge the postal workers who asked me to write this book, and I acknowledge the courage of those who agreed to endure the emotional process of their interview. Without sharing their personal experiences, America might never know the abuses suffered by many injured postal workers.

A very special thanks to my sister-in-law, Margie, a postal supervisor, who was instrumental in bringing significant meaning to role of postal management.

I am indebted to my business partner and long-time friend, Linda White-Andrews, who not only believes in my dreams, and adamantly encourages me to pursue them, but is always there to support me during those pursuits.

To Bobby Parker, my book cover artist, and Maria S. Karkucinski my book cover designer who worked long, and hard, to bring about my vision.

To my sister, friend and mentor, Dolores Belgrave who encourages me to work to my fullest creative ability, and advises me to always be fair to others. To my brother-in-law, Vernon Rivers, who encouraged, and supported me through the writing of this book. To Doris Johnson, my sister, and fellow author, and brother-in-law, Bob Johnson, who have always believed in my ability to achieve, and supported my every challenge. And to other family members, Ronnie, Margo, Bobbie, Stephanie, Kevin and Gregory, who labored with me to bring meaning to this challenge.

To my long time friends, Marilyn Jackson, Joan Pratt, and Michelle Towson who stand by me, and enthusiastically support my challenges.

And to my colleagues, Karen M. Singleton, Christopher Eddins, and Robert D. Carmignani who, on a day-to-day basis, gave me love and encouragement as I wrote *Out In The Open*....

CONTENTS

Foreword	xiii
Introduction	xvii

CHAPTER

I. Sergio—*They don't want me out of the post office...they want me dead* — 1

II. Cynthia—*I start remembering the sexual abuse* — 34

III. Megan—*I brought my egg and cheese bagel I know what mouse s--- looks like* — 41

IV. Dr. Wright—*I've worked with people ... the post office seems to be the worst* — 54

V. Caitlin—*I could smell the liquor on his breath...he pulled down my pants* — 72

VI. Valerie—*He kicked me in the stomach... I ended up in the hospital* — 91

VII. Dr. Tai—*I feel like the government is more insidious* — 120

VIII. Arnie—*I can see why people went to the post office and blew away a supervisor* — 129

IX. Gabriele—*I was soon to understand the reasons for such violence* — 138

X. Afterwords—*Is Anyone Listening?* — 155

Bibliography — 177

About the Author — 179

FORWARD

Out In the Open ... The Abuse of Injured Postal Workers is a compilation of interviews with injured postal workers who talk about their personal experiences with management abuse. Management abuse is endemic in the workplace, and permeates both private and government sectors. It is a national rather than regional problem. In some cases management is the direct cause of employee's injuries or have at least, in some way, contributed to their injury. In the federal sector management receives an annual bonus for maintaining a low percentage of work-related injuries. Considering this dollar incentive, federal employees believe that management will frequently lose, destroy, or intentionally misplace their paperwork in an attempt to prevent or frustrate the employee from filing valid workers' compensation claims. Federal regulation imposes penalties for this behavior, such as a fine of not more than $500 or imprisonment for not more than one year, or both; however, the regulation goes unenforced. Injured workers from the north, south, east and mid-west painfully illustrate in this book their experiences with abusive practices by management and federal agencies, which by law are obligated to provide benefits for bona fide work-related injuries.

Out In The Open ... also details interviews with health professionals who discuss their views and theory concerning managerial abuse of injured employees.

Out In The Open ... targets America's injured workers from both the private and federal workplaces. Both postal and private sector employees will either identify with the personal experiences of the employees in *Out In The Open ...* because they themselves have been injured, or they will be empathetic because they work with employees who have been injured and who are victims of management abuse. While the interviews may appear incredulous to some, others will know, "This is real." Whether federal or private all readers will find upsetting the interview with the worker beaten and repeatedly kicked in the stomach by a doctor to whom she was delivering mail. The next day, as she lay in the hospital with tubes in her stomach, her supervisor called and asked when she was returning to work. Readers will be alarmed by the interview with the postal supervisor raped on postal property and subsequently ridiculed by her supervisor when she returned to work. And readers will be astonished by the interview with postal supervisor who chose to defend a female employee who was continually harassed by the postmaster. This supervisor provided proper postal authorities with documentation of racist induced incidents by this postmaster. It is this supervisor's belief that higher officials retaliated against him by initiating efforts to ruin his career. In his interview, Postal supervisor tells of the continuing and on-going retaliatory incidents that led to the loss of his family, house, transportation, livelihood and health. This supervisor suffered a compensable emotional reaction for which the Office of Workers' Compensation Program refuses to pay benefits. This supervisor admits that he now lives in fear for his life.

America's workers who have never sustained a job related injury will find it frightening to learn that such flagrant abuse is America's dirty little secret.

Out In The Open ... is a portrait of contemporary and controversial issues relevant to workplace abuse, harassment and retaliation that potentially lead to violence in the workplace. It is a portrait of women and men battling mental health, politics and humanitarian issues, all stemming from management abuse. *Out In The Open* ... is a blueprint for health professionals who treat injured workers with emotional problems induced by management-related abuse. They will become familiar with the different *types* of abuse for which injured workers seek psychiatric and psychological treatment. *Out In The Open* ... focuses on the depth of the psychological impact caused by management abuse.

Out In The Open ... allows the readers to experience but a sampling of workers whose lives were ruined when they went to serve their government and became casualties of war in the workplace.

Out In The Open ... is not intended to slur, defame, or besmirch the character or reputation of any supervisor, manager, or higher official. The sole intent of *Out In The Open* ... is to bring widespread attention to abusive behavior by management that ruin lives. It is the hope of the employees interviewed in this book that government officials, and even Congress, will initiate systemic change to avoid continued violence in the workplace as a result of management's abusive behavior toward injured workers.

                                              Teresa J. Hill
                                              Attorney

August 1997

INTRODUCTION

When the postal workers in Oklahoma and California went on a shooting spree killing supervisors and co-workers, America asked, "Why?" When employers force injured workers to return to work before they heal, America's employees ask, "Why?" Injured workers who are harassed and abused by management because of a job-related injury ask, "Why?"

Every day, in private and government sectors, thousands of employees sustain work-related injuries. Some suffer emotional injuries such as anxiety attacks, while others sustain physical ones such as broken limbs. Many employees suffer harassment and retaliation when they file a compensation claim or engage in federally protected activity, such as testifying at an Equal Employment Opportunity (EEO) hearing. This is abuse.

Out In the Open ...is a compilation of interviews with injured workers who tell their story in their own words. The abuse of injured workers crosses gender, racial and geographical lines. The workers in *Out In the Open* ... are women, men, Caucasian, African-American, Hispanic, and Jewish injured workers who represent only a sampling of America's injured workers who were abused by their managers.

But there are many stories you will not read about such as the two postal employees who were threatened by co-workers and of management's failure to protect them. In one case a female employee was threatened by her male co-worker who threatened to shoot her. Instead of management taking action against the threatening male employee, the female, who was the *victim*, received a "letter of warning for creating a disturbance on the workroom floor." The male was transferred to a facility of his choice. This is abuse. You won't read her story because every story can't be told. There is the female employee who was a supervisor and suffered a job-related injury after a series of events. As a direct result of continuing physical threats from her subordinates and abuse from her co-workers and humiliation from management, she eventually experienced an emotional collapse. You will not read her story either because she is so traumatized and suffers anxiety attacks when she talks about her experiences with the postal service. Today, she periodically returns to the hospital as a result of anxiety attacks triggered by thoughts of her experience with the postal service. She can no longer work. You will also not read about the appalling behavior of the manager of the worker who severed his finger while delivering mail into a mail slot. As he was in the emergency room waiting to be admitted so the doctor could reattach the finger, his supervisor said, "... you can do whatever you want, as long as you report to work tomorrow at 7:30 a.m. or lose your job..." This employee is currently fighting a Notice of Removal because his treating physician will not medically release him to slot mail.

In some instances, overt violence toward management is an employee's only response. Frequently, however, innocent family members become the victims of the employee's rage. This was evident in the killings in Oklahoma and California where postal employees went on a shooting spree killing members of their own families.[1] Other employees' families are affected in other ways when their benefits are disallowed; they face loss of their homes, become homeless, or go on welfare. This was the case of one mail handler who thought his job was secure after working with the post office for a number of years. He had a spouse, two beautiful children, and a lovely apartment. One morning while lifting a heavy mail sack he injured his back; however, the agency terminated him as a result of the injury. The supervisor told the Office of Workers Compensation Programs that the employee said, [*I*]*n my country*

---

[1] The Time Mirror Company, *Los Angeles Times*, August 10, 1989, Part A; Part 1; Column 6: Metro Desk.

the women walk on the man's back to ease the pain ... that's what my wife did for me last night. Further investigation showed the postal supervisor had lied. This employee believed there was an underlying mission by management to rid that particular facility of people of color by harassing them until they either quit or transferred to another facility. Just prior to that employee's incident, three supervisors in the same facility assaulted another defenseless male on the workroom floor. He was then fired. This employee was not so lucky in getting his job back. Today, he is seen rummaging through garbage cans and mumbling to himself.

The Occupational Safety and Health Agency (OSHA) is always a step away from the workplace trying to insure that America's workers have a safe work environment. Postal employees feel that management covers up safety hazards by failing to report the true number of job-related accidents. One postal worker tells of the OSHA representative who came into her postal facility while she was working at the light duty table.[2] There were about 25 injured workers at the light duty table. The OSHA representative said he only had reports on "two" cases of job-related injuries.

In the postal service, job-related injuries is an economic issue since management is paid an annual bonus for a low percentage of job-related injuries. This sometimes results in management violating federal law by persuading or coercing injured workers not to file a claim. This happened with the employee who severed his finger while delivering mail into a mail slot. His supervisor told him if he would say the injury happened off the job, he would give him light duty status and let him keep his job. The employee reported this misconduct to the union, however, according to the employee, the union has not been helpful.

---

[2] Light duty assignment table for injured employees.

Frequently, injured workers receive incorrect claim forms which delay filing their claim. According to many employees, claims deliberately become misplaced, mis-filed, or simply are not submitted to OWCP for processing. *Out In The Open...* presents only a sampling of cases where such abuse has occurred.

While media reports of postal workplace violence have not identified the perpetrators as abused or injured workers, many postal workers believe that job-related harassment is, indeed, a factor. In her interview Cynthia states, *"I'm surprised at the person who has the patience to stand around and take the harassment over and over and over again ... one day you're going to just click ..."*

On August 20, 1986, a part-time letter carrier walked into a post office in Edmond, Oklahoma and fatally wounded 15 postal workers and injured seven others before killing himself.[3] On December 15, 1988, in New Orleans, Louisiana a six-year postal worker wounded three employees and took hostages. On August 10, 1989, a disgruntled 27-year letter carrier went on a shooting spree in Escondido, California. He killed his spouse and then went to the post office where he worked and killed three co-workers. A worker described the shooter: *"He didn't have any emotion. He was stern-faced ..."* The letter carrier then shot himself.[4] On October 1, 1989, *The Washington Post* reported that a 47-year-old Louisville, Kentucky man entered his former worksite armed with five guns "looking for bosses." He wounded 13 and killed seven before turning a pistol on himself. He was not a postal worker;[5] however, on June 26, 1994, the media reported that the growing violence in the workplace was due to harassment, threats, attacks, and employee stress levels.[6]

---

[3] *Los Angeles Times*, Late Final Edition (8/20/96)
[4] *San Diego Union Tribune* (8/10/89)
[5] *The Washington Post*, Sunday Final Edition (10/1/89)
[6] *The Columbia*, Sunday Edition (6/26/94)

While most newspaper articles provide excellent information about workplace violence, the reports do not indicate that the employees were injured or abused workers. However, postal workers interviewed in *Out In The Open* ... expressed their beliefs that if the shooters worked at the post office, there is the likelihood that they suffered some sort of harassment from management. It is their belief that management abuse fueled by home pressures can be the "straw that breaks the camel's back" leading to violence in the workplace.

In her interview, Alison talks about ways in which frustrated and disgruntled injured workers get their revenge on abusive supervisors other than going on a shooting spree. She said, *"I had a hand injury but was refused accommodation to light duty until I healed. My supervisor gave me a direct order to unload an eighteen wheeler (of mail) alone. After I had done so, he told me that he made a mistake; it was the wrong truck. He made me reload it."* She said her rage was so great toward her supervisor that she wanted to "blow him away" but thought about her small children and about spending her life behind bars. She decided it was not worth it. Instead, she confined her rage to mental thoughts of destruction. When she saw another worker taking tea to this particular supervisor she thought how nice it would be to *"tinkle in his tea."* She said, " *...the real revenge would have been to actually do it ... "*

This is a mild response from a woman who had been sexually abused as a child by her mother's various lovers since she was six years old, and then by male supervisors at the post office. The price she has paid for being a postal employee is by no means a small price. Hospitalized on 13 separate occasions for job-related emotional stress since being hired by the post office, she is again on stress-related leave.

Sergio summed up his experience: " ... *the postal service did win; personally they destroyed me, financially they destroyed me, professionally they destroyed me, physically and mentally they destroyed me... they completed their circle ...they don't want me out of the postal service, they want me dead ..."*

In order to protect the identity and privacy of the employee's interviewed in this book, and others who were not, names, identifying characteristics and other details have been changed.

Out In The Open ... The Abuse of Injured Postal Workers is powerful, riveting, and heartbreaking. The interviews you are about to read are America's dirty little secrets.

                                                           Angela V. Greene
                                                               Author

CHAPTER I

"She got to the point where she told me, and this was in front of her child, that she might go home and blow her brains out or that she wanted to blow the postmaster's brains out ..." Sergio

The problems started before I even got to this particular post office in a particular mid-western state. When I accepted the position, the person who was a supervisor from another post office explained to me that I was walking into a situation that I would not believe, and I wouldn't be able to do anything about it. He explained to me what was going on between a black female employee, and the postmaster, who is a white male. I listened, as I always do to what anybody says, but I have always believed in observing, and then getting my own opinion of a situation. I did listen to him; I took what he said under advisement, and I went to work with some insight of what was going on. As the weeks progressed, I found that the supervisor from the other office was exactly right. The woman he was talking about was the only black female, and the only black clerk in that office at that time. Being an ex-police officer, I've been subjected to many things, and I've seen many things during that time; seen many tragic things. But I don't think I was ready for the harassment and retaliation that I saw from the postmaster toward this particular female employee. It was embarrassing to me as a person, and certainly put me in a bad position as a supervisor because she was my employee. But the things he did to her, the way he talked to her, the way he treated her, it was demoralizing. It was very degrading. It not only affected me, it affected the people around this particular employee, even though they were white. But this didn't bother the postmaster. I talked to him on many occasions about what he was doing.

I was an up and coming supervisor, and I had to either stand up for what was right, or turn my back and let it be, to go up the ladder, so to speak. So, I decided that I would be a buffer, that I would stay between the postmaster and the female employee as much as possible, and that I would treat her the way she should be treated. I decided that I would try to keep them apart so that he couldn't treat her any worse than he was already treating her. But it didn't work out that way. I ended up going on detail to another post office. But during the year that I was in that particular office, every day it was evident that the harassment and the retaliation were very, very hard on the female employee, and very hard on the other employees in the office. It got to a point where the entire office was taking the heat for what the postmaster had felt against and was doing to this employee.

I've seen this postmaster do some outrageous things to her. While three or four white clerks would be sitting on a bench throwing mail, and she'd be right there with them throwing mail, he'd slap crackers out of her hand, and the white clerks would be sitting there eating sandwiches. I've seen him pour out her drink when the white clerks would be sitting, and drinking their drinks. When the postmaster walked into the building there are three things he would do every morning: he adjusted the thermostat, because I'd always turn it down or turn it up, and he adjusted the radio, because I'd always turn it down or turn it up, and he went straight to her. He was always telling her how stupid she was, what an idiot she was, and how sorry she was, and this just wasn't true. But this was his way of destroying her morale and self-esteem.

I remember one occasion when he was timing her. She was the only employee in the office that he timed on everything. From the time he walked into the building he timed her on everything she did. And on this particular occasion, I was in the back with them working and she had gone to the restroom and come back. Well, there was five minutes that didn't show up on the postmaster's time sheet, so he went out to her and asked her what happened during those five minutes. And she said, "I went to the restroom." He got mad and said, "Well, you retrack your steps." So when he did that I kind of walked with him. We walked back to the restroom and when we stopped at the restroom door, she said, "I went into the restroom and went to the restroom and then I came back out." Well he demanded to know what she did inside the restroom. At that point I stopped him and I said, "You know you're out of line, that's personal business and you don't need to know that." He got upset with me but he dropped it at that and stormed off. I told her to go back to her case and do her mail. This is just some of the things. So many things that he did. I mean, he'd try to get her for mixed box mail. And then a white clerk would tell him, "I'm the one that put these forms in the box this morning." Then his attitude would change and he'd say, "That's okay, just don't do it again." But if he thought it was her and he thought he could have something else to get on to her about, he would go for it. He would do this time after time again. She'd do a section in the box section and box mail and he'd come up to her and start yelling at her about it, and then come to find out she had left that section and another person had come to that section, and the fact is she had never put a piece of mail in that section at all. It was just sad.

I got to the point where I hated to come to work. I felt like there's nothing technically I could do for her. I had told the postmaster many times that what he was doing was wrong. I had talked to the proper authorities including the EEO counselor. On one occasion, the EEO counselor came down to the floor about a complaint that this particular employee had, and the EEO counselor asked me straight out if the postmaster was harassing this lady. And I said without a doubt, the postmaster was harassing the employee and something needed to be done. Well, I was told at that point that this thing is between this particular female employee and the postmaster and that it was going on before I got there; that it was none of my business, and that I needed to keep my mouth shut, close my eyes, turn my head, and not bother with it; that it wasn't my problem. Well, I couldn't do that. But, I ended up going on a detail which took me out of the office. I never returned to that particular post office.

Personally, I was glad that I was out of the office for a while because it gave me some time to get the stress out of me and become a normal person again. I had gotten to the point where I was constantly listening to everyone complain about what was going on with this particular employee, and how the postmaster was treating her, and it just got to the point where it consumed who I was, basically. And I took a lot of it home to my wife. Early on, she understood what was going on and she felt sorry for this woman, as I did, and as most of the people in the office did. But this situation eventually ruined my marriage. My wife left me.

When I left that post office, and I knew this was going to happen, the postmaster put another employee in my slot who was from the same office and who was going to do exactly what the postmaster told her to do. This employee did exactly what the postmaster told her to do, and together they harassed this particular black female employee unmercifully. The reason I knew about it all was because, even though I was in another office, at first, the postmaster would call me at home and he would brag about the fact that they'd given her a letter of warning that day. Then he'd called me a couple of weeks later and tell me that they gave her another letter of warning. And when he'd tell me what the letter of warning was for I'd tell him, "You can't do that. There's no postal regulation against what she did, and there's no standards to what you're saying." Of course, he just ignored me. This thing just escalated and got to the point where I knew she couldn't take it anymore, she just couldn't.

    The female employee came to my office several times, and on one occasion, she brought her daughter with her. Her daughter is probably 13 or 14 years-old. I could tell that she was upset. She sat down, she talked to me about what the postmaster was doing to her, and I knew that what she was telling me was true because I had observed it. I had seen it, and there's not a doubt in my mind that what she was telling me was an accurate account of what he was doing to her. On one occasion she told me that she had gotten to the point that she didn't know if she, and this was in front of her child, that she might go home and blow her brains out or that she wanted to blow the postmaster's brains out. At that time I knew something had to be done. I begged the higher officials to do something about it, and I told them it's time somebody needed to do something before somebody got hurt, whether it be the postmaster, or somebody else in the post office, or even this particular employee. You know, it's not a pretty thing for somebody to sit there and say they are going to blow their brains out in front of their child. So I decided at that time to write a letter to a higher official. I had thought about it for awhile and I had pretty much decided to do this anyway.

    A few days after she left my office, I got called to a meeting with the postmaster and another higher official. When I arrived at the office, I was really upset because I felt like I was being put on the spot. I was detailed into another office as an Officer In Charge (OIC), and I was not at that particular office any more, and had not been at that office for a number of months. So when I got there, the official had about 8 or 9 letters written by employees from that office. Each letter talked about the treatment of this particular black female employee by the postmaster and how degrading and disgusting they felt it was the way the postmaster was treating her. They talked about the atmosphere, and that it was one in which they felt somebody was going to get hurt. The official handed me the letters to read. The postmaster was sitting behind his desk. I read the letters, and he asked me what I thought of them. I, of course, already had a copy of each of the letters because the employees of that office trusted me, and even though I was out of the office on detail, they had sent me a copy of each of the letters. So I explained that I thought the letters were correct; I felt like somebody was going to get hurt, either the postmaster or the female employee, and God knows who else in the office, and that something should be done. The postmaster got hostile. He got very upset. At one point, he pointed to the other official and told him that the best thing that he could do was not mess with him. The postmaster is a very political person; he has a lot of political pull, and you just don't go against somebody like that. Personally, I would have fired the guy on the spot, but the official just seemed taken aback about it. We talked a few more minutes, and they explained to me that I needed to come back to the office the next morning, and go to the people and tell them that the postmaster had changed, and that he wasn't going to harass this particular employee anymore, and that the postmaster was going to be a model postmaster, and that the harassment was going to stop. Well, I told them I couldn't do that. This was the third time we had such a meeting. This was the third time he told me that the postmaster was going to change, and the postmaster hadn't changed yet; I told him that I didn't believe that the postmaster could change, that his hostility and hate toward this particular employee wasn't just for her as a person, it was because she was black. I explained to them that I would not, under any circumstances, come back to that office and lie to those people.

Before this meeting, on another occasion, they had another team come in to ask questions about the conditions of the office. Only it was the postmaster's friends who came in, so they asked the employees what the postmaster wanted them to ask. They really got nothing out of the situation. The employees already knew that it was a sham. They knew that they had been set up by the system that was supposed to be honest with them. And I wasn't going to come in and turn around and do the same thing to them. So I told them I wouldn't do it. The higher official got very upset with me, and he ordered me out of the office. He told me I could not return to that office under any condition. (Even though that was my office of origin).

Before I left, the official explained to me that another higher official had ordered him to come and see what was going on and to see if the letters were true. Well, he explained to me that he was going back and telling that higher official that the letters weren't true and that nothing was going on there. But this was a lie, a blatant lie. And the situation was getting worse. At that point, I had made my decision that I was going to write to the another high official. I was going to put together a package explaining what was going on.

THE PACKAGE

"The official explained to me that management of the postal service is like a fraternity ... you protect that fraternity at all costs ... the official explained that they were going to ruin my career ... and make an example of me."   Sergio

I wrote this letter which ended up being a package of documents. I sent it to the proper authorities. Ever since I've been in the postal service, I've kept a diary. So I sent over 100 pages of the diary (out of over 1,000 pages), to the proper authorities about what went on in the office where I documented incidents between the postmaster and this particular employee. I put the package together thinking, *"...Any person looking at this package would have to say there's a problem here."* All of the officials received their package. That same week, I got a call telling me that I had to go to a meeting with some higher officials to discuss the package.

I was under the impression that we were going to talk about the problems in the post office between the female employee, and what the postmaster was doing to her. I hoped that they were going to take the postmaster out of the office, or take her out of the office. I didn't care who they took out of the office, they just needed to separate the problem. In all honesty, they should have taken the postmaster out because he was very prejudice. He didn't like blacks; he made no bones about it. It had gotten to the point where it was very dangerous and I felt like somebody was going to get hurt. So I didn't really care which one they took out, they just needed to take one of the two out. And I thought this was what we were going to talk about.

The higher officials met with me. The meeting was started by telling me about protocol. I was told that district officials didn't appreciate the fact that I wrote to the other authorities instead of letting the district officials handle this matter. I explained that I had gone through all the steps that the postal service has put down. I went through the chain of command, but it had gotten to the point that there wasn't time to wait to see what somebody did. Something needed to be done now because there was the potential for violence. I told them that in this mid-western district the managers were not doing anything about the problem between the postmaster and this particular employee. I told them that it was my perception that they weren't doing anything about it.

They were very upset with me. There could be no doubt in anybody's mind that they were pissed at me, certainly not in my mind. They began to tell me that mangers don't go against fellow managers. They said to me, "Mister, there's a fence down the middle of this desk." There was a huge desk there. They said, "Let me explain to you what I'm talking about. There's a fence down the middle of this desk. The managers are on one side of this desk, and you have the employees on the other side of this desk." They said, "Management does not go across this fence, and if they do, they can't come back." It was explained to me that I had crossed the fence by telling on a fellow manager. I explained that I didn't see a fence in the postal service. I explained that I didn't care; it didn't matter to me whether it was a manger, an employee, or what. The situation had gotten out of hand, the retaliation and harassment against this particular employee had become unbearable, and that it was our fault. It was management's fault; we did this to this lady.

But all they cared about was that what I did was against the unwritten principles of management.

Then it was explained to me that management of the postal service is like a fraternity. I was told, "You protect that fraternity at all costs." They said, "It's not important what this person did; it's important that that person is protected by you." I didn't buy that philosophy, and that I totally disagreed with everything that was being said. It was explained to me that they were going to ruin my career. In fact, the words to me were, "We are going to make an example out of you," because they didn't want another manager doing what I did. It was clear to me that the officials present totally believed in this theory. At some point during the meeting one official said that they would see to it that the postal service ruined me personally. They explained to me that I had broken the unwritten rule: "You don't tell on another manager."

At that point I had little doubt that my career in the postal service was over. But I knew I had done nothing wrong. I had done what, ethically, I should have done; that which was morally right, and what the postal service says that they want its employees to do. They say they won't tolerate harassment; they say the won't tolerate retaliation in any form, no matter who it is. So even though what I did was right, I was going to be punished for this. I gave these officials credit, though. They looked me straight in the face, told me they were going to ruin me, and as events go along, you'll see they did exactly what they said they were going to do. It's scary that somebody can look at you and say, "You know, I'm going to destroy who you are." When I left the meeting that day, I knew that my career with the postal service was over. And that was hard to swallow. You know, I had done nothing wrong.

After that meeting I received a letter ordering me to go to an interpersonal skills class. This to me was very humiliating and very degrading since anyone who would read any of my recommendations, my merit evaluations, letters of commendation, or anything having to do with me and the postal service had always said that I had people skills; that I had interpersonal skills. They were doing this because they knew it was the kiss of death for a manager in the postal service to be sent to an interpersonal skills class.

Up to this point I had gotten outstanding merit evaluations from one of the higher officials at that meeting. That official sent everywhere. I was sent to different facilities as the trouble shooter, anywhere they had problems — they sent me to put out the fires. They had people calling me asking me how to do this, or how to do that; I was just doing all the right things for a person to move up in the postal service. And I had not been disrespectful, or out of line to anyone, any manager, anyone. I had done my job; I was proud of the job I'd done; I had done an outstanding job. In fact, when I left that particular post office, the postmaster had written me an outstanding evaluation. Also, the workers in that post office had written a poem about me. This was unusual because you don't find many federal employees that care enough about their managers to sit down and write a poem about that individual, and about what a caring, understanding person he is. But everything I had done to this point was outstanding as a manager, and as an employee in the postal service. The evaluations, the recommendations were all outstanding; I don't mean *good*, I mean they were all *outstanding*. But this stopped the day of that meeting.

THE DECLINE OF MY CAREER

"For several hours I met with higher officials, during which time they tried to get me to recant what I had written to the proper authorities...they threatened me with my job if I didn't. I refused." Sergio

I refused to go back on what I had said, and what I had observed and what had taken place. So at that point there was no doubt that these people were after me; they were going to be after me. And its extremely hard to accept that I simply told the truth and here I am now, don't have a career anymore with the postal service. Up to this point, I loved what I did. I mean I enjoyed being with people; I enjoyed working with the employees, and I treated them with respect and with dignity. In every office that I had been in, we worked as a team; we had the trust and the mutual respect between us. However, after the meetings, events started happening pretty rapidly.

Shortly after the meeting with the officials, one of the officials that threatened to ruin my career and ruin me personally, called me at the office where I was the OIC and told me that I could not return to my post office of origin. I explained to him that

that office was my office of origin, and that I had put in for that office and I didn't have any intentions of leaving. My family was there; we were happy there, my stepson was in school there. When he started school there he was a straight "A" student.

I tried to explain that I had applied for the available postmaster position where I had been detailed. Under the restructuring, I was entitled and qualified to apply for the job. At that point it was explained to me that I did not understand what the higher officials had told me in the meeting—that they were going to ruin me and they didn't care about the cost. We got in kind of a disagreement over what was said, and the remark to me was that they would "spend a million dollars to ruin me" This official then laughed and said, "You know what's interesting? Nobody can stop me." And it has come to pass, that's exactly right, because nobody can stop them.

It was explained to me that I *was* going to leave and if I didn't pick a place, a place would be picked for me and they guaranteed me that I would regret what was picked. Knowing that I was pushing the point of termination with these people, I couldn't argue too much. I knew that I couldn't stop what they were going to do to me. So, out of the places that were given to me to chose from, reluctantly, against my wishes, I selected another post office.

On that particular day, I found it hard to believe that nobody in the postal service would turn their head, that nobody would stop it. It was so outrageous. They cut my assignment and moved me to the new post office.

During this period of time, things at home started getting hard because my focus was on what was going on with the postal service situation concerning The Package I sent to the higher authorities, and now upper management was threatening to ruin my career. Up to that point, my wife tried to be understanding, and she was. But when I went home and told her we had to leave, that was very upsetting to her. That was a hard thing for her to take because we were happy, you know. We were a happy family. We had ties in the community. And now, I was uprooting the family, the child in school, everything, simply because I told the truth. That was a hard thing for her to take. Eventually, that's what led to our marriage ending in a divorce, because of this post office situation.

I arrived at the new post office and much to my surprise, the postmaster was none other than one of the officials who earlier wanted me to retract my statement about the problems at the other post office. This was an obvious embarrassment, because at one time this official presided over more than 100 associate offices and now was the manager, and postmaster, of just one office.

Upon my arrival, it was explained to me the displeasure with me being there; I was blamed for the postmaster's — demotion, so to speak — so I was assigned to work with the clerk craft. The postmaster was so busy trying to get the office straight, and I was so busy trying to stay out of the way, that there really wasn't the opportunity for the postmaster to start the harassment and retaliation toward me that I believe they really wanted to do.

One day, however, the postmaster called me into the office and informed me that there was a notice saying that I was due to testify in the EEO hearing of the black female employee. It was explained to me that if I testified I would be destroyed as a manager. Well, they were already working on that. I was asked if I was going to testify, and I explained that I had all intentions of testifying if I was called. Immediately after that meeting the postmaster began coming into the office early in the morning and harassing me. I mean, arriving early almost everyday, and argue with me over volumes; telling me I was falsifying volumes, falsifying different reports; that I wasn't doing my job on schedule. I've been doing this sort of thing for many years, and I've never had anybody question me in all this time. The harassment had escalated to the point where I had had enough.

In 1995, there was a mid-year review scheduled for the postmaster to attend. Well, the crew of higher officials was going to be there. Now, these higher officials had already told me that they were going to ruin my career, and at this point they were seeing to it. But the postmaster came up to me and told me that I was going to be required to go to this meeting. I specifically asked that I not be made to go because I knew I was going to be harassed, and I said that. I was told that I would go, whether I liked it or not. In an office meeting just prior to the up-coming meeting with the higher officials, I asked again to be excused from attending the meeting because they were only going to harass me. I stated that I had not been in that office for the first part of the fiscal year, and I knew nothing about the operation. And again, I was told that I would attend.

Well, I attended the meeting along with the postmaster and other officials from different facilities. As we sat there, the presiding official sat at the end of a very long desk surrounded by the other officials. One official went through the city delivery, and the city delivery operation was like 2000 miles over plan and was running terrible. They simply looked at the supervisor from our post office, and when the supervisor told the officials that they'd try to do better, the officials said: "Okay, nothing big." Then they moved to discuss the clerk craft. When they got to the clerk craft, the manager of operations support explained to me that I ran the worse mail processing unit in the mid-western district, and the worse clerk craft in the district. Well, I had not been there for the first four months of this fiscal year; however, this craft was running 68 hours under plan as we were sitting in that meeting. Therefore, the statement was totally in error. I explained that we were 68 hours under plan, however, I was told that I didn't know what I was talking about and when an argument ensued. The post office operations manager (who does the district finances) finally stopped this official and said, "Look Sergio's right, he's under budget coming in today." Well, they got into a small disagreement. But I *was* 68 hours under plan.

The management of operations asked me would I save 14 hours a day and I said "no" I wouldn't; it was physically impossible to do. I was asking for *more* hours in my budget because I knew the operation could not continue to save hours. At that point an official bolted out of a chair and asked me if I knew what the heck I was talking about? That I was guaranteeing additional hours and here I was asking for more hours than were in the budget. I felt the hatred for me. And the fact that this is one of the officials that said they were going to ruin my career, this was their way of showing their peers and their managers that they were not happy with me. And you know as the saying goes, "...if that official is not happy with you, neither is anybody else." So when this particular official sat down, I explained that their statement was in error; that I did not guarantee additional hours a day, that I could not. I told them that what I said was that I would work within the budget that I set up, which was the addition of more hours. And you could tell that this official was visibly angry with me, and that the events of the earlier meeting had not subsided at all. I became very upset, very humiliated, and very embarrassed, and I felt like maybe I was two or three inches tall. It's like, you're sitting there in front of managers who have your career in their hands. *They* make the decisions on your promotions, and *they* were there at that table. This official had just destroyed what little bit of hope I had left of going anywhere in the postal system. So I was angry. It was visible that I was angry. And it was visible that the official was angry.

I left the meeting that day very upset, I was upset all the way home. But the harassment didn't stop. The postmaster immediately came back to the office threatening me about this, and threatening me about that. I'd had enough. The postmaster had come in one day and threatened to fire me on several occasions about things they thought that I had done. The next day they came in again and threatened to fire me about several things they thought that I had done. Then, when they found out that they were wrong, there wasn't even an apology. On one occasion they just remarked, "...lucky guess," and walked off. You know, you just get to the point where you have had enough. You don't care what it is, or who it is, and I had reached that point.

EEO COMPLAINTS AND EMOTIONAL STRESS

"I spent Ninety-nine percent of my time trying to keep my job, to stay focused on what they were doing to me...in the process, I neglected my wife and my stepson and I had what people call, 'tunnel vision'...I ate, slept, drank, thought, and lived the postal service..." **Sergio**

I called the EEO[1] office and talked to the head EEO counselor. She could tell I was very upset. I explained to her that I wanted to file an EEO complaint for harassment and retaliation against the postmaster, and other higher officials for what they had done to me during the meeting. I told her that I had documented every thing the postmaster had done, what was said, and what went on. But the EEO counselor's concern was not that I wanted to file an EEO complaint. She could tell that I was very, very upset, very distraught, very anxious. In fact, she, on several occasions, tried to

---

[1] Equal Employment Opportunity

calm me down and suggested that I see an EAP[2] counselor and get some help. She explained to me that I was entitled to 12 sessions free and that she would make the call and take care of it for me, but that I needed somebody because she could tell that my health was in jeopardy.

So that's what she did. Matter of fact, it wasn't 15 minutes later when I got a call from the EAP coordinator. I talked to him and he could tell I was very upset, very anxious, and he immediately called the EAP coordinator counselor who immediately called me. We talked, I don't know, maybe for 30, 45 minutes or better, and he set me up with an appointment which I kept. This was the first time I had seen the man, and, very honestly, I couldn't trust him. In fact, that's what I wrote in my diary that day. I felt I couldn't trust the man, but at that point I went because I knew I needed to get some relief, some help.

Two days later the postmaster came in early again and began threatening me again, and when we had our regular supervisor's meeting later that day my job was again threatened. I was told that if I couldn't do my job, I would be fired. I was very upset. During the meeting, the postmaster made a few other remarks like, I needed to "go back to craft", and this was being said in front of a person who was just an acting supervisor, and in front of a craft person, as well as a fellow supervisor. I was humiliated and became very anxious because what was being said was totally non-acceptable to me; it was embarrassing, especially for other people to hear.

We got on the subject of a grievance that I had already settled, one that the postmaster ordered me to deny. I tried explaining that I had already sustained the grievance and contractually, I could not deny it. Well, an argument ensued. I mean a big argument. in fact, the postmaster got so upset that I was abruptly told, "clock off, leave the building." The postmaster then

---

[2] Employee Assistance Program

got up and left. Right after the postmaster left the other supervisor who was present apologized to me for the postmaster's action, because the postmaster was very abusive toward me and it was very embarrassing for other supervisors to hear this coming from the postmaster. They could tell I was upset, very upset. I began having chest pains and I felt like they were severe. I felt like if I could get away from the office, because I was so anxious, so upset, that they would go away. Well, I went home and laid in the bed.

During this time, I spent ninety-nine percent of my time trying to keep my job, to stay focused on what they were doing to me. I would keep notes and try to fight a system that is unstoppable. In the process, I neglected my wife and my stepson, and I had what people call, tunnel vision—I only saw one thing, that what they were doing to me was wrong, and somehow I had to get a handle on it. I ate, slept, drank, thought, lived the postal service, and what they were doing to me. I don't think there was ever a time during this period of time, that I ever thought about anything else. I certainly neglected my family, which ended up causing my divorce. But unless a person goes through this sort of thing they will never understand what it's like. You become what you are consumed with, and in my case it was the postal service. You become that; that's what you become; you don't become anything else; you sit home and you think about what they're doing to you, you think about the dishonesty of the system; how powerful the system is to intentionally pawn you. And nobody can stop it. I mean, it just becomes who you are.

I became depressed. It got to the point where I wouldn't sleep at night; wouldn't sleep during the day; wouldn't sleep at no time. If I slept, I slept four hours, I'd wake up 25, 30 times during that four hours and the only thing on my mind was the postal service. I sat and watched TV, and while looking at TV I would think about the postal service. I quit working out, I stopped lifting weights; stopped going to the gym. I bought a computer to try to keep up with everything that was going on, and it just became everything; my being was trying to overcome what these people were doing to me because I knew it was wrong. I was totally consumed by it. I mean people don't understand what it is until it actually happens to you. It's far more than you disagreeing with somebody; it's far more than you doing something wrong and somebody coming up and firing you. It's the injustice, it's the inadequacy, and the loss or lack of justice in the system that makes you ... I don't know, it consumes everything about you. And that's what it did to me.

During this time, my wife and I became very far apart, and it was me. She never knew one day to the next whether I had a job, how long I was going to have a job. My wife expected me to feed the family, which I did, to have a house, which we had, have automobiles, which we had, make payments, which we did. But she got to the point where she felt she couldn't be with me any further in that situation, and *I couldn't leave it.* So she left me. And trust me, you can't leave that situation. You can't walk away from it, you can't do it. I mean, I *wanted* to do it. During one of my sessions with the EAP counselor, I told him that. I said, "I want to leave. I want to walk away from it. I want to just get away from this thing, but I can't do it." And you can't. You cannot, you cannot walk away from a situation like this.

I went back to work, but even in the parking lot my chest started pounding as I approached the building. I became very anxious; I had to make myself go to a place I once loved to work, the United States Postal Service. And the more I got to thinking about what the postal officials did to me, the way it was done, the way I was humiliated, the more anxious I became; and the more anxious I became, the stronger my chest pains became. So I called another supervisor to come in and when she got there, I left. Before leaving, I filled out a CA-1[3]. On the form I marked the box "continuation of pay," and I left a note for the postmaster asking to let me go see a doctor. I knew at this point that they were after me

---

[3] Traumatic injury workers' compensation form for job-related injuries

so bad that if I went to see a doctor on my own they wouldn't pay for it. And financially at this point, I couldn't afford to go on my own. I had owed, from an earlier accident, a substantial amount of money, and even though it wasn't my fault, the insurance company had not paid yet. Therefore I owed hospital bills. In my state, you can be refused emergency room attention, or hospital admission if you owe hospital bills. And I've already been humiliated and embarrassed enough by the postal service, I didn't think I could stand going to the hospital in front of everybody, and them looking at me and telling me 'there's the door,' you know, 'hit it.'

So I went home and waited for the postmaster to call me. The call never came. I called three separate times that day, and the postmaster refused to talk with me. In the meantime, I called the EAP counselor, and he set up an appointment for the next day, and he talked to me probably 30, 45 minutes. I went to his office the next day, and he suggested that I take time off immediately. I had annual leave left.

According to the postal regulations, you can take emergency annual leave, and upon return you need to give proper documentation. Well, I requested an emergency leave. I did that up front because I knew that the postmaster, and the postal service, hated me so bad that my situation wasn't going to be the norm, but rather I was going to be the exception to the rule. I took the EAP's letter to the postmaster, and the request for the hours of leave, and I went home. I went back to bed. When the postmaster saw it he sent me a letter denying my annual leave, and carried me AWOL. Now, I was more anxious, more depressed, and more confused because any other person, and many have in that office, got emergency annual leave for things like, they had to work on a car; things like, their brother broke his arm so they had to stay with him for two weeks, you know, stuff like this. But here I was having chest pains, I was in severe medical distress, depressed, very anxious, and the postmaster tells me I'm going to be carried as AWOL because they didn't like what the EAP counselor wrote.

I went back to the EAP counselor and he suggested that I find a psychiatrist because I needed medical help in the way of medication, very badly. So I sought the services of a psychiatrist. He immediately put me on three types of medication, and he put me out of work for another two weeks. I applied for advanced sick leave because I had no sick leave left, and under postal regulations you're entitled to advanced sick leave with a doctor's note. Well, I gave the doctor's note to the postmaster and requested advanced sick leave. I was turned down even though I still had 80 hours of annual leave left. I ended up calling a higher official, and he eventually got me my two weeks annual leave, but he couldn't overrule the postmaster giving me the advanced sick leave. The postmaster also refused to allow me to go to a medical doctor as I had requested, which is also in the postal regulations.

Things were really on a roll now. There was one retaliation case after another. In one incident the supervisor who was the postmaster's gopher started rumors about me having affairs with people on the workroom floor. Plus, she never filled out the CA-2a[4] that I gave to her for processing. In fact, I found it on her desk a couple of months later. She had not signed or submitted either my CA-1, CA-2, or CA-2a. I'm sitting, and waiting, and thinking that OWCP[5] was going to help me out somehow, but they didn't even have my forms. My claims were never sent to them for processing.

In another incident of retaliation, the postmaster informed me that they were stopping the postal service from paying for my continued EAP visits, even though I still had visits left. I was told that if I wanted to go to EAP I would have to pay for counseling myself.

I learned that the postal injury compensation specialist had altered a document by transposing her name onto OWCP letterhead. With this fraudulent document she was able to get

---

[4] Recurrence of a job related injury

[5] Office of Workers' Compensation Programs

medical information about me. She also called my psychiatrist's office trying to get my medical information from him, but he caught her and explained to her that she posed as a service provider for a medical corporation that I supposedly had insurance with. He told her that he would not release any information to her.

In another incident, I went into work one day and found this same supervisor going through my files. She had no business reading my files, she was not a "need to know" person. I found her invasion of my privacy humiliating and degrading. I mean, my file had a lot of medical information in it; it had a lot of information from my psychiatrist that she had no business reading. I filed several grievances on the postmaster, and on the supervisor because of their continuing, and ongoing harassment and retaliation against me.

Things really started heating up. The postmaster began placing immense pressure on me, harassing me, finding reasons just to bring me into the office, one on one, to threaten to fire me when nobody could hear. I had gotten to the point I couldn't take it any more. I contacted my psychiatrist who immediately put me out of work for 45 days. Three days after returning to work the postmaster had me back in the office. This time, I was told that they were going to conduct an investigative interview with me concerning performance issues, and this could lead to my removal from the postal service. Well, I got upset about this and called my attorney who in turn called my workers' compensation representative, who, in turn tracked down my psychiatrist about 7:00 p.m. that night for an emergency conference. I drove for an hour and a half to his office for an emergency session.

I gave the postmaster a letter requesting that the investigative interview be postponed until my attorneys could be present. The request was denied. I asked that my treating physician be allowed to accompany me. That request was denied as well. There was nobody else that I could trust at that point. I couldn't trust anyone in the postal service because I was a target. I was a prize for anyone that could get me. I knew at this point that the postal service had made me a target, and whoever got this target would be rewarded for it.

On the day of this "so-called" interview, I walked into the meeting and requested that the door stay open. I wanted to have the conversation with the door open. The postmaster ordered me to close the door. I did. They brought in an employee legal representative from another office which was 130 miles away, even though they had a legal representative much closer. They began asking me questions, but the questions were not about performance issues as I was told, the questions were about "The Package" I had written to higher officials concerning the postmaster from the first post office, and the black female employee who he continually harassed. They were asking me questions about the EEO cases that were coming up. I got upset. I began having chest pains and I explained to them that I needed to leave. I had medication out in my locker that I could take, and I needed to get away and take some medication. They refused on three separate occasions to let me leave under the threat that if I got up and left that they'd take disciplinary actions against me.

For almost two hours straight, one person asked me a question, another person asked me a question; they were just asking me questions one right after the other. It got to the point I began having severe chest pains, and I got up to leave. I was going to leave no matter what happened. At that point the postmaster told me to sit down, and the legal representative advised that I be given "the letter." It scared me because I knew these people wanted to fire me, and I figured that the letter was telling me that they were terminating me from the postal service. I read it. It was a letter giving me my rights to that investigative interview. They gave me the letter almost two hours after the interview started, and pointed out the very first paragraph which stated that my refusal to fully cooperate in an investigative interview could lead to my removal from the postal service. The postmaster then told me that if I didn't sit down and cooperate he would fire me.

This was an interrogation; it was not an investigative interview for performance issues at all, and that incensed me. I became so upset that my chest pains became more severe. At that point, I asked to be taken to the hospital. The postmaster refused to dial 911 and advised me that they would take me to the hospital. I was driven the distance to the hospital in silence. The postmaster took me to the hospital and dropped me off in front of the hospital emergency room, and when I say 'drop me off,' I mean that's just what happened. I got out of the car, and they drove off and left me.

THE POSTAL SERVICE WON
THEY HAVE DESTROYED ME

"They were taking more than $380 a month, in addition to over $180 they cut from my pay...it was costing me $160 to drive to work...the postal service did win...they personally destroyed me, financially they destroyed me, professionally they destroyed me, physically and mentally they destroyed me, they completed their circle..." **Sergio**

They downgraded me, and made me drive to a distant post office, but the harassment didn't stop when I got there. The agency tried to collect for the continuation of pay I had received because the workers' compensation claim I filed was denied. But I was still waiting on the hearing. I explained to them that I had not gotten the hearing yet, and asked if they would wait until the hearing was over. They told me that they would not wait. When they told me that, I did something that most people in the postal service don't even know about. If there's a collection action against you in the postal service you have the right to ask for a federal Judge to intercede. Ninety-nine percent of the people in the postal service don't know this.

So I filed a request in Washington to have a federal Judge take this over. A federal Judge wrote back and said, "I am Judge so-and-so, and I will be handling this case." The postal service immediately got me into the office and said, "Look, we'll give you a year's grace, and we won't do anything until your OWCP cases are finished." You know, all that time I had been worried about what I was going to do, how I was going to live because they were taking about $380 a month away from me, in addition to the $180 they had cut from my pay and the fact that I was now paying an average of $120 to $160 a month for driving back and forth to work, which was 85 miles round trip, I was in financial trouble. I was at the bare minimum of what I could do. I couldn't even afford my house any longer. I had borrowed money, and people had helped me. Friends had helped me, and I no longer could ask for their help or expect their help, so I had to file for bankruptcy. I thought at that point they had destroyed me as a person, as a human being, that it would be over, but it wasn't over.

I think one of the most humiliating things for me, of all of this, was having to stand in front of a court room full of people while an administrator talked about my bankruptcy, what I lost, and have the creditors come up and take away from me what I had. That along with everything else was a very humiliating experience for me. I lost everything, I lost my home, I had to leave my home, I lost my automobiles, I lost my credit cards, I lost everything.

I signed the agreement with the agency thinking that they would honor it. Boy, was I mistaken. They'd already done everything to me—everything I think anybody could ever do to somebody. But I thought what I had left was to look forward to at least having a Christmas. Well, in November 1996 the agency began taking the $380 a month out of my salary, even though we had signed an agreement that they wouldn't do this. When I called the employee labor representative about it, he said, "We'll take care of it. It's wrong; we're going to take care of it." Well, I went through December with no Christmas, without anything, without any money, in fact, had it not been for my brother buying me food and stuff like that, I'd been in a bind. He lent me the money to buy groceries and pay my bills because I couldn't afford to do it. I didn't buy anybody any presents; my daughters didn't get any presents, my stepson didn't get any presents, and I didn't get any presents from my family, so I didn't have a Christmas. But after Christmas, in February 1997, the agency stopped taking the $380 from my paycheck.

So the postal service did win. They did win. They destroyed me, and my family; they destroyed my career, they destroyed me professionally, they destroyed me financially, they destroyed me mentally, they destroyed me physically. They had completed their circle. They did what they said they were going to do, and nobody stopped them. I mean, you know, I was not even the same person, and don't think I will ever be the same person that I was before all this happened, before I sent "The Package" to the proper authority.

THEY DON'T WANT ME OUT OF THE
POST OFFICE—THEY WANT ME DEAD

"How do I live knowing these people hate me so bad..."
Sergio

In 1997, I had a hearing relative to one of my cases. At that point, I thought the postal service had done all they could do to me; humiliate, embarrass, and ridicule me as much as they could. I was wrong. During the hearing, in front of a federal Judge, the agency representative tells this Judge, my attorney, me, and the court reporter, that I could have applied for a position in Oklahoma City; that they would have gladly sent me to Oklahoma City or California. You know, federal people died in Oklahoma City. That's well known. What a lot of people might not have known is that there's been three separate incidents of killings by postal employees in California. So this agency official was telling me that they want me dead. They don't want me out of the postal service; they want me dead. How do I live knowing that these people hate me so bad that they want me dead? That's, you know — I still don't truly believe that someone could say something so malicious like this and get away with it. And he meant it. He was serious. He was as serious as anything; I mean, he was dead serious.

So now I have to deal with the fact that these people want me dead. I would not, after all that I have been through, I would not put it past the postal service to hire somebody to harm me. I mean they've called the police department on me, they've lied on me, they've tried to catch me doing things, they destroyed me in every way they can. What's left to do but to kill me? To put somebody on me to hurt me? I mean, now I have to live with that everyday; wonder if I go to work today is somebody going to be out in the parking lot to blow me away; am I going to walk out of work and somebody will shoot me, or is somebody going to shoot me at work? Will somebody rig my ignition so when I turn the key my car will blow up. You know, that's what I have to live with.

What they've done to me will never be erased. I feel like maybe somebody is listening, and seeing what these people are doing to me.

When the agency got the decision, they sent me back to the post office because a Judge ordered them to send me back. I reported back to work, and you know, it's like I don't even exist. I'm what they *call persona non grata* in the postal service; I don't exist. I don't even have keys to the building. Here I am a supervisor, supposedly supervising people, but I don't even have a key to the building. I don't have access to any of the rooms in the building. I don't have access to any of the vaults, or the security code to the alarm system. I can't make any decisions. I just don't exist. I just plain don't exist.

So, here I am today since this all began, never knowing from one day to the next whether I have a job or not. I'm wondering, you know, am I going to get hurt, is somebody going to hurt me? What's going to happen to me? You know, can't nobody stop this. And while most people would say, "Why don't you just walk away from it," I can't walk away. I can't. *I cannot walk away from this.* I know I can't, I can't let myself just walk away. I mean, the mental aspect of this, of the last two — three years, I don't think people would have to be in my shoes to understand mentally what a person goes through. I mean, I have said on countless occasions, sat at my house and you know, I thought about killing myself because...[6]

You file workers' compensation claims, and they're good claims, but since the OWCP is in the same building as the postal service the agency takes your claims, and *they*, not OWCP, *they* answer the claims. So workers' compensation doesn't have any say in it, basically. OWCP is *told* what to write, or they just rubber stamp what the agency writes. You know this because when you read their decisions, their decisions are so outrageously stupid that you know the postal service is in bed with workers' compensation. It's absurd; anybody can see it. There's no reasonable person that could read any of the OWCP decisions about me and say that they're fair, and honest decisions. And this is the system that you have to live within.

What people don't understand about the postal service is they're so powerful, they're so strong; hell they don't answer to anyone. There's *nobody* that the postal service answers to. You can't call the Inspector General's office and get them to handle anything against the postal service. You can't call a Congressman, or a Senator and get them to do anything with the postal service. It can't be done; they're not going to do it.

What a lot of people don't know is, and it's been my personal experience, through this experience, that the postal service employee cannot win an EEO at the first step if the postal service does not want you to win. They can't win because the postal service writes their own decisions. If a Judge finds in your favor, which in some of my cases Judges have found in my favor, the agency then will say that the Judge is stupid; that he or she does not know what they're talking about. But, if the Judge finds against me, which has also happened in some of my cases, the agency will

---

[6] *Sergio is emotional at this point. He pauses before he continues*

rubber stamp *that* Judges decision and say, "Hey, the Judge is a genius, he's right." And the sad part about it is, like my first case in front of a particular Judge in the New York District, he had my promotion cases, and it took him almost a year after he heard the cases to make a decision. Then when he made a decision, his decision isn't even talking about *my* cases. He doesn't even have the information right. But because he found against me, the postal service rubber stamped it and said, "Yeah, this man's smart, and we agree with him." You know, the dishonesty permeates the postal service. You can't even call the postal inspectors because even they can't be trusted to do the right thing.

During the course of one of my cases, my attorney, and myself requested documentation through discovery. The postal service would not cooperate, but the Judge finally had to order the postal service to send us these documents. When they finally sent the documents, we found out that the postal service had done a background check on me, like the CIA would do on somebody. They documented where my family lives, where my children went to school, where their mother works, where their grandfather works, where they live, how much child support I pay, where my children were raised. They documented how many times I've been married, what the number is on my marriage license, when I was divorced, who I was married to, how many kids I got. They even documented when my mother and father were divorced.

They can tell you they legally got my personnel file from when I worked as a police officer many, many years ago. That has absolutely nothing to do with anything I could ever do in the postal service. My family, and the fact that I pay child support, could never have anything to do with what I do in the postal service. Everything I did while I was a police officer has been outstanding. It's well documented, it's in my personnel file through letters of recommendations, and commendations. Yet, they went to the police headquarters, lied to them and told them that I was under investigation for criminal activity, and got my personnel file, trying to get something on me.

You know, this isn't an honest agency, it's not a fair agency. When you get an agency this powerful, this strong, something needs to be done, because I know can't nobody stop what's going to happen to me. Whatever happens to me can't be stopped. I know that the system, the postal service, is too powerful. Nobody's going to stop this.

CHAPTER II

"It was all about sexual harassment ... he was very, very explicit with his words ... he touched me several times and made me feel very, very uncomfortable ... I wanted to crawl under a rock because I didn't know who to talk to ... he came behind me and pushed me against the wall and said, "You don't know what you're missing if you don't come with me ..." Cynthia..

    I started working at a post office in the south in December of '86. I had mild arthritis in my wrists and I worked as an LSM [1]clerk for two hours a night. Then I transferred to a post office in the east in July of 1989. I did the same LSM work there only, it was for twelve hours, sometimes thirteen, fourteen hours a night, sometimes seven days a week. It was crazy. When I walked in there I said, "No way. We have to do all this? We didn't do this at the other post office." But hey, you know, it was my job so I did it.

    From the minute I walked into that post office, I felt uncomfortable. All the eyes felt like they were just staring me down; the men as well as the women. I just wanted to turn around

---

[1] Letter Sorting Machine

and leave and go back home, but I had voluntarily transferred, and I wanted to do what I went there to do, and that was to do my job. But it was difficult for me for the first several months that I was there.

One day I was sitting casing mail[2] and some guy came over and asked me a question. Apparently one of the supervisors, which was a lady, became upset at that. She came over to me and told me and him to get back to our seats. The next thing I know she's sending me to another post office. I had just moved to the area and did not know my way around. I asked her to give me directions to the other post office where she wanted me to work, and she gave me the wrong directions on purpose. I would think that's harassment. She knew exactly where to send me, why send me all over the state? And that's just the first incident. Things kept happening after that, and it just went on and on, and it did not stop.

In another incident, my supervisor, a male, started harassing me sexually. He made remarks like, "I want to touch your butt," or "Let's go out and have some lunch," or "Boy, would you just give me a night?" I was very, very uncomfortable. I ignored him. I did not know what to do. I was new there; I just wanted to do my job; I didn't want to be fired. So I continued to do my job, and ignored him. He actually tried to fire me because, I think, in my opinion, he wanted to sleep with me, and I did not want to do that. There were many incidents after that with the men there.

There was another guy, who also sexually harassed me, going as far as coming to my home and knocking on my door. My children opened the door, and I couldn't believe my eyes. I said, "What are you doing here? You're my boss." You know, it was all about sexual harassment. He was very, very explicit with his words. He was fresh. He touched me several times and made me feel very, very uncomfortable. I wanted to crawl under a rock and hide because I didn't know who to talk to. I was so afraid of losing my

---

[2] Throwing letters into mail cases

job, but I would not give him the satisfaction to let him touch me. I would pull away and say, "How can you say things like that?" At the time, I did not know who to turn to, where to go, who to talk to. There's not a lot of support in the post office for things like this.

This kept going on for a long time. There was another male, a 204B[3] who also sexually harassed me. One day I was behind the LSM sorting the mail and he came behind where I was and pushed me against the wall, and said, "You don't know what you're missing if you don't come with me." Prior to that incident, he had been harassing me all along. He went by my house, he sent me flowers and stuff like that. All of this was just driving me crazy. I was like, "Why is this going on, what's wrong with these people here?" I did tell a supervisor what was going on, but nothing was ever officially done about it. Eventually, the guy transferred and went to some other place. I was relieved, thinking that I wasn't going to have to look behind my back all the time now. But it just kept going on and on, because when he left there was someone else to take his place.

In spite of the sexual harassment, for two and a half years I did very well, until my wrists started hurting and swelling up. I had a lot of inflammation with pain. I went to the doctors, and they told me my arthritis had become severe, and that it was aggravated from the repetitive motion of keying on the LSM, and casing mail. I went on light duty for awhile and I thought I was getting better, but I didn't.

There was a male supervisor who was harassing me during the time when I was on light duty. In the beginning, I would think of him as a friend, and at one point, I had confidence in him. I would go talk to him and tell him my problems, and what was going on with the post office. But he took advantage of our friendship. He took my confidence too far. He said to me, "I know you're on light duty because of your wrists, and your arthritis, but does it hurt

---

[3] Acting Supervisor

when you have sex?" And that really shocked me. I mean I got up from the light duty table, and I just wanted to throw something at him. I just wanted to wring his neck. I ran to the bathroom in tears.

There was nothing I could do. I just felt like I was pinned against the wall the whole time, from the minute I walked into that post office until the day I left. There was really nothing I could do. It made me feel miserable; it made me feel like I was worthless. It made me feel like I wanted to crawl under a rock because I did not know who to talk to, or where to turn.

It wasn't like I could go to the union and ask them for help, because they act like they don't even believe what's going on. They favor the supervisors and upper management, not us little people that have to do the harder work. And it's crazy.

The post office is just a crazy place. I never knew what I was walking into when I was walking into it. Working next to these people is just like being in a hell hole; I just wouldn't want to go back there – it's scary emotionally, and physically it drains you. You just have no support whatsoever from anyone there. At least, I never found any support.

I had a car accident, and they sent me to the doctor and he checked out my hands. He said, "My goodness, you have severe arthritis, you need some operations. You know your wrists are deteriorated." He said, "At this point you're not going to be able to do anything if you don't get the operations." So I did that. I had three operations in 1991 and one in 1992. All along, I was going back and forth to work between the operations and doing light duty work. Management treated me like pure crap because, I guess, once you work as a light duty person you're not worth anything any more.

There was one female supervisor who would come by the light duty table and tell all of us that we were "not worth a paycheck," or better yet, we were worth taking outside and killed "like they do to horses" when they're crippled. We heard this, and different other things continuously; and we could not do anything about it. We just had to sit there, and just look at her like she's crazy, and just take all the stuff she threw at us. That really hurt. We asked ourselves "Why, why would she say that to us?" We *were* worth something, but she made us feel like nothing, and it was just, crazy.

There's a lot of violence in the post office. I've seen supervisors' cars get destroyed. What they do is they put acid over the car so the engine can blow. I've seen a supervisor get slashed in the face with a knife. I've seen men beat up other supervisors, whether it's a female or a male; and that's only because of all the harassment they take; and some people just can't take it. I mean, there's a point in time when you think about it, and you say, "I'm human, treat me like a person, not like an animal." All this just continues to go on, and nobody cares.

I've heard of the Oklahoma and California killings, and frankly my comment on that is I can see why. They're pushed so hard to the limit. I'm surprised at the person who has the patience to stand around and take the harassment over, and over, and over again. One day you're going to just click. They, management, drive you insane. They drive you to the point where you just want to give up on everything, on yourself, on your family. It's not only the job that it affects. It affects your family at home. You just, you're not functioning anymore, you snap at your kids, and it's all because of stress. All the thinking you have inside of you --- 'Are they going to do this, what are they going to do next; am I going to be able to go there, do my job and just leave?' No. It never happened that way for me. It was just one thing after the other, and I just wanted to know when all of this would stop?

Eventually I filed a CA-2[4] for workman's compensation, and it took two tries before I even received any information about my claim. A lot of my files were getting lost. I was wondering why I would always go up to the front desk and ask for a certain file here, or a certain file there, or a copy of this and a copy of that, and

---

[4] Workers' Compensation Claim for Occupational Illness or Disease

they would either push me away, shove me off, or say "tomorrow, tomorrow, tomorrow". And when I finally got the copy, it was the wrong copy. I just felt like I was being jerked around, or that some of my papers were being lost purposely. I think this is true, because the papers were there one day, and then gone the next. That's crazy.

The people in the postal injury compensation unit, they all kept delaying my stuff. I think they delayed everything to the point where they wanted me to be tired, and to go away, or wait so long that I would give up so that they could say, "Yeah, she did it, you know, she did that on her own; we didn't do it to her." But by them doing all this delaying, I still didn't give up. I kept going back.

The post office terminated me. They said that I was unable to perform the job that I was hired for, even though I have documents from several doctors supporting the fact that the aggravation of my rheumatoid arthritis was in fact, caused by the repetitive motion of my job.

After I was fired, I received the letter of acceptance of my claim from OWCP for aggravation of my rheumatoid arthritis, but that was a year and a half after I was terminated. It's crazy. I thought that was really odd for them to do that. What is OWCP doing with my records? What is the post office trying to do? Do they think that everyone can be manipulated or intimidated? I mean, some people might be, but some people just won't give up. And I will not give up my rights to everything that is mine, because I have worked hard for it.

When I received the acceptance letters from OWCP, a year and a half later, they turned around and denied it again. I don't understand what's going on. Something that was accepted, was denied a year later; how could that be? Nothing in my medical condition changed. My rheumatoid was still there, and it's permanent. They were jerking me around and I didn't like it. I wish there was someone, somewhere that can help us postal employees fight for our rights, because we do have rights, and we are human beings.

After I was terminated from the post office, I felt lost. I did not know what to do. So I went back home to be with my family, and to have their support. But I was miserable and stressed out. I didn't know what to do, didn't know who to turn to. I felt so lost because the post office, and OWCP, turned my life around. Even with friends, I lost all my friends, I lost my self-confidence, I lost everything.

The reason that I'm giving this interview is so that people out there could know how postal employees are treated. I want everyone to read this, and think about how in-humane the postal management can be with its employees. I consider the abuse from the post office to be tremendously harmful to everyone. From the minute I walked in there, it was abuse 'til the moment I left that place, there was abuse. From the moment I walked into the post office, I was sexually abused by management. How much can a person take?

I am now in school trying to get my life back together, and my emotions intact. I'm trying to forget all the abuse that I went through from the post office. Even now, when I have hearings and trials and things like that, the OWCP hearings and the EEOC[5] hearings, things that I'm fighting for that I have the right to fight for because of what the post office did to me physically and emotionally.

I start remembering all the sexual abuse ... I almost feel like I'm there at this very moment, and it makes me cry, it makes me angry, it makes me want to go back and stand up and say "You cannot do this to me anymore, no more..."

[At this point, Cynthia breaks down and cries uncontrollably, unable to continue the interview.]

---

[5] Equal Employment Opportunity Commission

CHAPTER III

"I went to the main cafeteria where they served hot food ... and this particular day I bought my egg and cheese bagel ... when I got down to a quarter of it I noted that there was something that struck me immediately as mouse turds ... it was like tunnel vision. I started to vomit ... I grew up dumping mouse turds out of my shoes in the morning, I grew up looking at them all over the stove, all over the kitchen table, everywhere, so I know what mouse sh-- looks like." Megan

When I began my postal, initially I thought that I had made the right decision. I was assigned to a post office on the west coast where I started my orientation, and my 90 day probation period. I left the bank where I was working and came to the postal service because I felt it would be an opportunity to advance myself, and have a good career. I liked doing payroll, and accounting, and that's mainly what I was interested in getting into once I got into the post office. Instead, what I found was a whole lot of double standards, and a lot of confusion, a lot of discrimination, a lot of who-you-know, not what-you-know, a lot of that. The incident with my first male supervisor created, to this day after ten years, a view of the post office that maybe the post office wasn't such a good idea after all. Maybe I should have gone into another industry.

Initially, I felt the need to respect my manager. I wanted to do the job right; I wanted to advance, and I wanted to satisfy management. Instead, what I found was a shock, because the type of behavior that my supervisor displayed just wasn't something that I was prepared for. I was expecting to be dealt with professionally, instead, this man, and his ways were abusive. I found that he was using his position, and his power for his own personal gains.

I first realized that there was a problem with management when I was told that you only get a half hour for lunch. But, once I started working, I found out they make up their own rules; you can really take an hour and a half lunch, depending on who you know. Then I realized I had a much bigger problem when in the middle of my 60-90 day probation the manager tried to get very friendly with me. A little bit too friendly. I am 5' 10", honey-colored brown, and I have shoulder length wavy hair. The manager offered to buy me lunch, he stared at me a lot, and his closeness made me feel uncomfortable. His actions were not professional. His behavior was inappropriate. Quite inappropriate. For example, he would get a telephone call and instead of taking it in his office, if the phone was near me, he would lean over me to answer the phone. He would brush up against me. I could feel his breath on me. Now that is close. I found out that he was supposed to be a Christian, and married, and I felt like I had nothing to worry about. I was wrong. It turned out to be a thing where he not only used his power, he abused it.

During my probation period, he would make little comments about my probation, and about my upcoming evaluation. One day I went to lunch, I knew I had a half hour, and I believe it was the day I was attempting to get involved in the credit union or something like that. I went with a male employee who knew where to go. When we went to the main building I saw that we were running late, so I figured that I better call and let the manager know that I would be a few minutes late. To my surprise, when I returned to the station I found that he was angry at me. I felt it was the type of anger that you would feel coming from someone that you know on a personal level, where you hurt them personally, not on a professional level where you did something against regulations.

If any of the male carriers showed too much attention to me, this supervisor would lash out at them, "get back to work," or lash out at me by placing me in areas where I was isolated, up in the back to stamp canceled letters and stuff like that. On this particular day, when I returned late, he took me into his office, he was very upset, and he said to me, "We're gonna to terminate you." Of course, I got real upset because I didn't want to be terminated. He started telling me that I was wrong for coming back from lunch late; it just seemed like he blew it out of proportion. He threatened me to the point where I started to cry. I started to beg for my job, I started to beg for his forgiveness. From that point on my employment with the post office took on a totally different meaning to me. I had only been there a couple of months and I didn't feel secure. I felt threatened, I didn't feel at ease. I felt that I had to be on my guard at all times.

One thing led to another. When he saw that I was going to resist his advances, he started to use his supervisors to do things to me. For instance, a money order in my batch mysteriously got lost. I explained to him that the money order was there when I tallied at the end. But it was missing, and I was held liable for it. And just every little thing I did. That's where the tension mainly started.

I later learned that he never submitted any paper work to have me terminated, it was all verbal. At that point I had already gone to EEO and filed a sexual harassment complaint naming him as the discriminatory official. They came in, they questioned me, they questioned whoever else they had to question. I also went to labor relations who told me that I could get a transfer to another post office on hardship. But sometimes I feel that EEO and labor relations run parallel to management. Otherwise, why would they give me a deal where I have to take a step down in my grade? Anyway, I was told that I would become a mail handler once I arrived at the new facility.

When I arrived at the new post office, I was met by a woman who was a reassignment clerk, and at that point I was expecting to be told that I was going to be a mail handler. Instead, she told me, "I don't know who told you that but there are no mail handler positions available at this time. There are only mail processing jobs available. You have fifteen minutes to decide if you want that." I felt desperate, I needed the job. I wanted to stay in the post office, and I knew I could not go back to the other post office which turned out to be an emotional hell for me. I took the job.

I was given the night tour, which was the first time I had ever worked nights. I attempted to adjust, I changed my dress code, since I was now working around machines. I went from corporate to jeans and T-shirts, and I attempted to do this job as a mail processor. They gave us orientation, the same orientation that I got when I initially came into the postal service. I felt that I was getting it twice. I got the idea that the orientation was set up to instill a threat, that they were saying, "Well, if you do this, this is what we will do." After what I had experienced up in the other post office with my former supervisor, I felt that orientation was definitely set up to intimidate, and to prepare you to be conditioned into an industry of discrimination, an industry of a lot of double standards, a lot of who-you-know, and not what-you-know.

As a mail processor, they didn't give me any training. Training came from the next person showing you the ropes, not from a person specifically trained to train you the proper, and safe way. They *spoke* about safety and the proper way to do a job, but they didn't show you *how* to do it. They were good at making presentations. The highlight of training was not to get your fingers cut off. The way the machines were made they were very hazardous. The moving parts on the machine were exposed.

INDUSTRIAL NOISE

Not too long after being transferred I was assigned to work with the machines. I learned that dealing with the noise was the major part of this mail processing job. Noise that you had to be able to work around like in an industrial environment. Since this was the first time for me, I found out very quickly that it wasn't physically, or emotionally good for me. The machines, the OCR[1] and the BCS[2] were the main machines that were on this floor. I started to feel very nervous. I couldn't understand why I was feeling so nervous. Why I felt the need to find some quiet place, you know.

I looked forward to the breaks. When I found that I could use the restroom as an option, I started to hide out. It was more like hiding out because I remember many times when I pushed that lady's room door open, and I closed it behind me, it was like a sigh of relief. The noise was gone. The ringing was still in my ears, but the noise was basically gone. I dreaded having to go back out onto the workfloor. It got to a point where I spent so much time in the lady's restroom that management started to page me. The

---

[1] Optical Character Reader
[2] Bar Code Sorter

supervisor would page me almost immediately because, after awhile, he knew exactly what was going on. I did bring to his attention the effect the machines were having on me, and his response was, "If you can't do the job, maybe you need to go somewhere else." That kind of closed me down. I felt helpless, I felt trapped. Working around the machines is difficult because you've got to be able to tolerate the noise that's already in your head, and then the noise that's around you from the machines.

In addition to the constant environmental noise, my depression stemmed from the interaction that I had with management, and their response to me. It had to be retaliation for filing that EEO on my former supervisor in the first facility where I worked, because I leaned that filing an EEO complaint on management is definitely a no-no. I've been told by co-workers that I've been blackballed because, "You just don't file EEO complaints against management." When you do, it goes in your folder and once it does, all of the managers know to watch out for you. You are branded as a troublemaker. You just don't file an EEO on management, because that affects them directly, which makes them very, very angry.

THIRTEEN HOSPITALIZATIONS

I started to feel anxious. My body had to adjust to nights as I attempted to work in this very noisy environment. I recently had a new baby, and I was up during the day. All these things made it difficult. But what made it even worse, I guess, was the response from my supervisors, and their insensitivity, "Well, if I can tolerate it, you should too," or they would say, "See, no one else is complaining," which made me feel like something was wrong with me; that maybe I can't keep up, or that I can't match up. I started to become very depressed. Feelings of failure started to surface, and I felt like I needed to put myself in a hospital. And that's what I did.

I experienced my first hospitalization as a result of work related-stress. While I was in the hospital they did a profile on me which showed that I was under a lot of stress, and that I was experiencing a lot of anxiety and depression. Treatment lasted for thirty days. Since then, I was hospitalized thirteen times in the ten years I've been with the postal service. All of the hospitalizations were job-related. Management's mistreatment led me deeper and deeper into depression.

I was put back to work the day after my hospitalization, right back in front of the machines. Three months later, I wound up signing myself into another hospital where I spent another thirty days. I was treated for depression and suicidal ideation's. I just was not able to deal with the noise. I asked management for an accommodation to an area away from the noise, but their response was that it was not possible to allow me to cross crafts. But at the same time, I observed other clerks in cross craft positions and nothing was said. I learned about one man who had been in a cross assignment for four years. But, of course, he was a white male.

I was totally overwhelmed and found myself very, very depressed. Of course, all of this filtered over into my personal life. I found that I wasn't as active, I didn't want to get involved in the usual things, my mind was spent just thinking of how I could stay out of management's way during the eight hours that I was at work, and how to get around the noise. I met a doctor who helped me to understand that the noise, and the anxiety. He explained that the type of interaction I was having with abusive management had a lot to do with why I couldn't tolerate the noise. He put me on medication, and requested that management transfer me to an area that didn't have a lot of noise. But I still experienced a lot of harassment. For instance, there was a tour superintendent whose idea of giving me an accommodation was placing me in a sub-basement, which was like a dungeon.

To add insult to injury, I learned that all of my medical documentation had been exposed, and circulated to all of management, which made me feel very, very uncomfortable; now everyone knows that I have "psychiatric problems." This lowered my morale and increased my depression and anxiety. The hospitalizations made me miss so much time from work that this particular tour superintendent felt it was necessary to recommend my removal. But it was rescinded because I supplied all of my medical documents. The documents had been there the whole time, but, no doubt, this was their harassment.

I took another downgrade and was transferred over to the Detach Mail Unit where there was no machinery. There, I encountered a supervisor, and a tour superintendent who had a personal relationship with one another. He called her 'mama' and it was very, very friendly. The supervisor was verbally abusive. He would tell dirty jokes, and expected you to laugh. He took advantage of the casual workers, the females mostly. He was dating a woman in the office, but that didn't stop him from making his advances in terms of his nasty, dirty talk, and stuff like that.

I showed him no interest, and one night he placed me on the platform loading 18 wheelers. He had me loading and unloading them by myself. The procedure is to have two to three people load and unload an 18 wheeler, but he had me unload it by myself. Once I unloaded it he told me that it was the wrong truck. There were moments when I felt such rage, I felt so angry, so violated, so walked on, and so worthless, that on this one occasion when he asked me to bring him some tea, I felt like tinkling in it.

At that point, I had another hospitalization for the depression. Just feeling totally worthless in the postal industry. I felt like I didn't have a purpose, that somewhere along the line I missed my purpose. Somewhere along the line, they missed my meaning for being there.

Eventually I was assigned to another facility as a mail handler, but again, with machinery noise. At that point, I requested a detail to another station, but the plant manager said they all they could give me was a custodial position. Again, my hands were tied, so I took what was available to me. I didn't like the idea of cleaning toilets. It was degrading. All the time I walk around with the feeling, "How come you people don't see that I am a qualified person; that I am capable; why are you still condemning me for taking action on something that I had a right to take action on?"[3]

It's like they want you to almost give up your rights to everything, and to submit yourself to any and everything. If they're having a bad day, hey, that's your tough luck. You just have to handle it. Drink about it, smoke about it, or whatever you have to do about it, but just make sure you are there to make the numbers match, and that those numbers tie out at the end of the tour. It all just creates an atmosphere of hostility, and a lot of rage, a lot of rage.

---

[3] Referring to filing an EEO complaint against management

THE CONTAMINATED BAGEL

One day, I went to the main cafeteria where they served hot food. Every morning I would order an egg and cheese bagel. I was on medication and needed to eat, so I would start my morning by coating my stomach with food. On this particular day, I bought my egg and cheese bagel. My co-worker and I went and sat at the breakout on the workfloor. I ate half of the bagel then went over to my building where I kept my medication. I took my medication and started to finish my bagel. When I got down to a quarter of it I noticed that there was something that struck me immediately as mouse turds and I became totally .... it was like I had tunnel vision. I felt sick to my stomach. I started to vomit. I felt, again, how could they allow this to happen to me?

The bagel was contaminated, and it had been sold at the private food vendor inside the postal service. I later learned, when I tried to sue them, that they didn't even have a health permit.

I showed the bagel to my co-workers, and then took it to a post office nurse who said that it might be burnt onions. I took it to the postal workers comp specialist and she said," Oh, it must not be that, it must be something else."

But I knew damned well what it was. I grew up dumping mouse turds out of my shoes in the morning. I grew up looking at them all over the stove, all over the kitchen table, everywhere, so I know what mouse sh-- looks like. There's no debate about it, burnt onion, sesame seeds, and mouse turds are very distinct in the way they look.

I felt humiliated because of the way management acted about it. They wanted to totally disregard that it ever happened, and they attempted to cover it up. The post office contraverted my workers compensation claim, and OWCP denied it. I'm still fighting that one; the post office is still taking 300 dollars out of my check behind that incident. [4]

To add insult to injury, I hired an attorney to handle this case. I trusted this man, but as it turned out, he had some serious problems of his own. I gave him the bagel because he was supposed to take it to be examined by some expert. The next I know, he called me and told me that he said he lost the bagel. *That* was the evidence. I could not believe it. *He lost the evidence.* Or so he said. The bottom line is I lost my case because there was no evidence to be examined, even though I had a medical reaction from eating the contaminated bagel. I know that what was on that bagel was mouse turds, not sesame seeds.

I was hospitalized as a result of that incident. I had stopped eating. I felt paranoid that I was being poisoned, and whatever I ate I thought something was in it that I missed. I just totally became turned off with food. It was very difficult, very difficult. And it caused me a lot of stress. I started to abuse myself in other ways to get over the feeling of hunger; and just the inadequate feelings that I felt as a result of management's response to the situation. Their insensitive response, and their denial that the event ever happened caused me a lot more stress, and a lot more of anxiety and tension.

My reason for giving this interview is so that other federal employees, specifically in the postal service, specifically women, and new people that are coming in, will know that if they're

---

[4] At the time of this interview, Megan was experiencing the pay deductions

experiencing anything like what I have, they should believe that it's real. They have to know that they're not alone.

I hope that perhaps one day Congress, or someone, will come in and see that postal management needs an overall change in their management system. They need to look at the people they select to be a part of management, and how to train management to boost the morale of the workers so that some person does not take a rifle, and go back there and clear out the office.

CHAPTER IV

"I've worked with people in the airline industry, hospitals, and other agencies, but the post office seems to be the worst ..."
                                                                Dr. Wright

This is my perception of the people that I've seen who work in the postal service, there seems to be an arbitrary abuse of power by postal management toward employees, especially employees who stand up to management, employees who are willing to challenge some of the practices that favor certain people.

It seems like these abuses of power by lower management are either covered up or supported by upper management. In other words, there's nothing done about these things that are documented by employees and witnesses. And what's interesting, and I'm sure you found this too, it doesn't seem to have, as far as I can tell, have much to do with race. Because the managers who do these things, abuse people, are for example, Caucasians abusing Caucasians, or African Americans who are abusing African American employees. So it's not always a racial thing. I mean, the higher up Caucasian management will support their lower level managers who may be or not be Caucasian. It's like a conspiracy. They stick together, regardless.

There's an arbitrariness about the way management deals with employees who they perceive to be problem employees for one reason or another. Usually, it will be people who stand up to management, especially after they file a grievance or a claim for workers' compensation; then management will go after the employee with a vengeance. I mean it's like a system which is designed to eliminate any challenge or question to their power, which is often quite arbitrary. It's surprising to see something like this in a federal agency like the post office.

You would think that it wouldn't be happening. You would think that maybe you would see that sort of thing in private industry. But this is worse than anything I've ever seen anywhere. And I've worked with people in the airline industry, hospitals, and other agencies, but the post office seems to be the worst. I mean there were people who had been harassed, sexually harassed in one way or another. Nothing's done about these people, the employers. That's the amazing thing. And the employees have to fight for their simple rights.

In one case, I treated a postal employee for severe emotional stress because after he had confided in his postmaster, they're both Hispanic, that he had been arrested, the postmaster went behind his back and repeated to other postal employees some of these confidential statements that the employee had made to him. I think this was in retaliation for the fact that he had filed a grievance because he didn't get a promotion to a position in a particular facility where they never had a Hispanic supervisor. The retaliation was apparent.

It doesn't really matter who you are; it's management against employees and they just stick together regardless of race, creed or color. So, it's really a system of corruption, and abuse, that is just really hard to comprehend in a federal agency like the post office.

In another case, I treated a female who also worked for the post office. Her postmaster had his own little band of favorite employees. Well, she worked on the night shift where most of the work was being done by a few female employees while the men would be sleeping, or goofing off, or going out for coffee or drinking, or whatever they were doing. And she protested. But you don't do that in the postal service. You don't do that; you do what you are told to do, you don't question it. But it was grossly unfair, and she's a gutsy kind of a kid, and she protested. She thought it wasn't fair that she and a few other women were doing all the work, and these guys were goofing off.

Well, the postmaster went after her, as did some of his cronies who were also supervisors below him. They started to harass her in one way or another to the point where she was still in her probationary period when the postmaster terminated her. But he terminated her after the deadline of her probationary period, so the termination was invalid, and she got her job back. This infuriated him, but it didn't stop him from going after her again. He said to her, "I'm going to get you, sooner or later, I'm going to get you." Then she really just started falling apart. Emotionally, she just could no longer handle it.

I've been dealing with psychological or emotional problems more than physical injuries, but that's worse in some ways. What happened with this particular woman was that she had a breakdown when, finally, as the last straw, she was working on the night shift with some of the postmaster's cronies and one of the male employees threatened her. He said something like, "I'm going to f------ kill you", or something like that, "I'm going to shoot you." Well, when she reported the incident to her supervisor that night she told him that she wanted it reported to the postal inspector, but he refused to do it. He should have done it immediately. So she reported the incident herself, which she should not have had to do.

The next day when she reported for work the postmaster writes her up for being disruptive rather than taking action against the guy who threatened her. She had a nervous breakdown. Her breakdown was not so much a reaction to the threat, but to the fact that instead of doing something about the employee who did the threatening, management went after her. She suffered from a post traumatic stress disorder. She just fell apart at the unfairness of it. And of course, underlying that was a tremendous amount of rage that this would happen to her. She was out for quite a while, about nine months or something like that.

When she was finally ready, I had requested from the postmaster a transfer to another post office until her safety could be assured, and of course, he blocked the transfer. He could easily have transferred her, but he blocked the transfer. I think because he wanted to maintain his power over her, and finally to force her out of the system. Get rid of her. From his point of view, she was a troublemaker simply for standing up for her rights. You know, this is the kind of thing you see. They don't want any questions asked about their arbitrary ways of running their little shop.

When it looked like she wasn't going to get her transfer, which would have been easy enough to do if he wanted to get rid of her, but no, he wanted to kick her out of the post office completely, I concluded that maybe she was ready to deal with whatever was going to happen. She had to return to work, she had a family, she needed the money. Then he started to harass her *and* me by questioning my judgment about her readiness to return to work. This went on for quite a while. He then sent her to the agency psychologist consultant who concluded that she was perfectly capable of returning to work. This was all so unnecessary.

It is my understanding that she wasn't even compensated through workers' comp for the time she was out of work for her job-related injury. She suffered post-traumatic stress from continuous harassment by the postmaster. Clearly, it was a work-related injury, that's why she was out. It had nothing to do with her family life, there were problems there, but that's not why she was out of work for nine months. Her family problems stemmed from the problems on the job. She was out of work because she was in shock that her manager would write her up for simply reporting that she was threatened by a male co-worker, who later was rewarded, by the way, by being transferred to another post office of his choice. But this is the kind of thing that goes on. That is just one example.

Then there was the other female postal worker who was working in this particular post office, where she was an excellent employee. She had been given supervisory responsibilities without the title or compensation, and as I recall, there was a change in postal practices, a reorganization. It was a speed up where she was being pressured to take on more and more responsibilities, almost to the breaking point. This is a woman who had never seen a psychologist up to that point, who was never on any kind of psychotopic medication until that point.

She had a very good work record, too, until she went into the office of her supervisor who verbally abused her in front of other employees, peers, and some supervisory staff. Then she went into shock. She went to the agency nurse right after it happened. She had heart palpitations, she had symptoms for hypertension and trauma, and the nurse sent her home. She then went to see her personal doctor who prescribed some psychotopic medication for her anxiety. Sometime after that she returned to work. When she did return to work she was challenged by some of the employees that she had been supervising. I believe her authority had been undermined by her supervisor when he badmouthed her in front of the employees. So she lost control of her employees, and she had a breakdown. She never returned to work after that. And that, again, was job-service related.

I believe it was probably directly triggered by that reorganization plan which put pressure on everybody. It was like a pressure cooker situation. It was almost like a sweat shop where they're working people harder and harder, giving them more and more responsibilities. One of the responsibilities she was given was as a safety officer where she would be taking people to the hospital for their various injuries. But then, her fellow supervisors would be angry at her because *they* had to take over her caseload, which was drastically increased. The number of people she had to supervise after that reorganization had an impact on her. Of course, I don't know about the impact of the reorganization of the postal service on other employees, because I only spoke to her. She was my patient as a result of the pressure of the reorganization. But the impact on her was like the sorcerers of penance, I mean she was just working herself to death.

It was after the reorganization that a lot of her stress symptoms really started to develop in a general way. Not specific to any particular abuse by any individual manager until there was that occasion I mentioned before when she was insulted, deeply insulted by her manager. But the beginning of that breakdown happened after the reorganization. You have a very clear cut before and after behavior pattern where prior to that, she didn't have any problems. She went to work, she didn't miss work, she never took time off from work, and she didn't use any medication for her high blood pressure until the reorganization. Then she was pushed, pushed, pushed to the breaking point, and the final straw was when her supervisor insulted her and she just fell apart.

She was humiliated, and it's amazing to me that in the evaluation of her workers' compensation claim, they wanted causal relationship. I wrote my report, but for some reason they didn't see these incidents at work as causally related. Here's a woman that was functioning okay, she's insulted by her supervisor, she goes to the nurse, the nurse finds her blood pressure elevated, she goes to her doctor the following morning, and he finds her blood pressure elevated. She does all of this right after the incident with the supervisor, immediately after that. And they denied the claim later saying that there is no causal relationship. Then about a month later when one of the employees that she was supervising basically refuses to comply with her request, she just really totally breaks down. She feels like she's lost her credibility. Again, she goes out of work the same day, in fact she fainted, as I remember. Then she goes out and never returns to work, and they're saying there's no causal relationship. What do they need to establish causal relationship? You almost think that they made up their mind beforehand that there is no causal relationship, regardless of what the evidence is.

So there's some kind of connection. I think workman's comp is geared to saying no, regardless of the industry, there's no doubt about it. But here, it seems to be like there's no willingness to support a claim. I guess maybe that's part of the problem when you're not dealing with physical injury, but I've understood that even with physical injuries they will challenge the claim.

This is a relatively young woman, she's in her forties. Why is she not working? She never went back to work. She was unable to return to work. So this was like permanent damage, I mean this is not a temporary thing. She has, still has, a sense of hopelessness, suicidal feelings. She's a happily married woman so it has nothing to do with her personal life. This has to do with the speed up, the reorganization plan, which is really a speed up, to get more productivity or whatever, to the breaking point. I don't know what the rates of breakdown are among other employees, but I understand there's a lot of pressure. It's like a pressure cooker. You hear about people going in there and just blowing up, and killing somebody from that pressure.

There's no doubt about it, she did not have a pre-existing condition. The only thing she might have is that she is a very sensitive person. But she's tough, I mean she was. She didn't go to her doctor because of the reorganization directly; she coped with it as best she could. It might have gotten to her sooner or later, but the fact is that she went to the nurse, and her doctor the following morning after she was insulted publicly by her manager. So it's not that she can't take it, but maybe she was vulnerable to that kind of abuse where she was treated with respect prior to that. She was given all kinds of responsibilities, and then she's treated this way by her supervisor in front of others, which clearly undermined her authority with the people she was supervising; this led to her final breakdown a month later.

My understanding of the eggshell theory is that people can have pre-existing conditions; some of them may simply be sensitive. In fact, the people that I'm seeing are sensitive people, maybe that's part of it. They're not thick-skinned people, they're sensitive people who are vulnerable to abuse. And I'm thinking of somebody like my patient who I spoke about earlier, whose supervisor betrayed his confidence, he is one of those people who is sensitive. In his case, this is a man with a great deal of dignity. He had a wife and a child, he was active in his community, and he was part of management at a post office. But he was naive enough to confide in his postmaster. He was honest, he wasn't trying to cover up anything. He was just telling the guy some things about himself and then this man, the postmaster, betrays that confidence by telling everybody.

I came to understand later that some of his medical records were shared with people who should not have seen them. So he was tremendously humiliated. And to add insult to injury, his application for a management position in a post office where they never had a Hispanic manager was denied. Instead of hiring him, they hired two white men, with clearly lesser qualifications. He couldn't help but believe that because of his race, he was denied that job. But whether the manager is or is not white, they stick together.

A manager can be just as abusive to a fellow African American, or Hispanic, as one who is white. It's not simply something like in the South where you find that kind of abuse. But postal management abuses regardless of the color of the manager or of the employee.

THE NEGRO BASEBALL LEAGUE

"He was Caucasian, he and his wife got involved with the Negro Baseball League ... he came in one day to the office wearing a jacket from the Negro Baseball League ... Well, so-and-so is not going to be too happy about that ..." Dr. Wright

It has been my experience that management may perpetuate job-related injuries. One example that comes to mind is one white male postal worker who I began to treat for severe depression. He's another one who is sensitive. You have people who, for whatever sensitivity they may have, are very sensitive people; but I wouldn't say so sensitive that they couldn't do their job. In other words, until they were abused by management they had good track records, they weren't missing work, they had good performance reports. I've come to see a before-and-after abuse. In other words, these people are not breaking down because they couldn't do their jobs.

This one man had a very good track record, but he had a supervisor who most people would say was a supervisor from hell. The supervisor must have had some kind of political connections or whatever, but he was very arbitrary. He could get away, literally, with murder and there was nothing that could be done about this man. Once he decided to get rid of somebody, that was it. I would say this is almost a general pattern with each of these people, that if they are determined that some employees are troublemakers, for whatever reason, they're going to do everything they can to get rid of them.

Anyway, he was Caucasian, and he and his wife got involved with the Negro Baseball League and he came in one day to the office wearing a jacket from the Negro Baseball League. There was some publicity about it, and one of his fellow employees said, "Well, you know, so-and-so is not going to be too happy about that." Nothing was ever said directly to him, but he was transferred to another post office where he was given responsibilities that he was not prepared for. He had been doing a good job where he had been trained.

What I remember is that he was given responsibility for money and stamps in the new post office. He was not trained for those responsibilities. He had seen an example of what happened to his predecessor, a female who was charged with loss of money or stamps that were in a safe area, and he was afraid to assume that responsibility which he wasn't trained for. Nor was he given any training; just thrown into this job for which he felt inadequate and unprepared. Not as a person, but one who simply was not trained to do this particular job. So in his case, the emotional damage was caused by putting him in a job which he was not trained to perform, and he felt very inadequate, and fearful of not only having to pay for any losses, and of ultimately losing his job. And so, he just fell apart. This guy, no doubt is a sensitive man who was doing a perfectly good job until he was being abused by his manager.

The manager confined him in an area of the post office where he had to sit with the safe. People would walk by and make fun of him as if he was a prisoner of some kind. They drew pictures of him. He felt almost claustrophobic in this space. But I think more than anything, he felt overwhelmed by the responsibility which he was not prepared to assume. There was no attempt to train this man to do this job, to prepare him for it. His predecessor apparently had some kind of a breakdown too, because she couldn't cope with it either. Now they throw this guy in there and you almost feel like he was being thrown to the wolves so that they could get rid of him. Why did they want to get rid of him? It could conceivably have happened because of the white postmaster's attitude toward him about his participation in the Negro League. It certainly was nothing about his work, and he wasn't the kind of guy who got sick.

These employees that we are talking about had a good track record as far as attendance; they weren't late, they weren't sick, they weren't "troublemakers", until they had a confrontation with their supervisor, for one reason or another, and then management went after them. It seems like the system was designed to just get rid of them.

THE DOWNGRADE TO TOILETS

"She kept getting downgraded and downgraded and downgraded ... she's being well paid for cleaning toilets, but it's humiliating for her to be doing this ... this is not what she wanted to do ... she worked for the post office for ten years and was hospitalized 13 times for job related stress ..." **Dr. Wright**

Another lady that I began to see was a good example of the eggshell theory. Here's a woman who was sexually abused at a very early age by her mother's various lovers. Yet, she went to school, she functioned, she tried to overcome her childhood horrors. She had a job, she had a good work record, and then she went to work for the post office. She was looking forward to working for the post office. She had some computer skills, and she thought, "Great, they'll use my computer skills." Next thing, she's working in a post office where her postmaster, also, starts to sexually harass her. This is a good example of the eggshell theory; she was vulnerable to that experience of sexual harassment because she had been sexually abused as a child. She should not have expected to be sexually harassed by her supervisor at the post office. Now, that led to the beginning of her breakdown because she challenged her supervisor's conduct. She filed a grievance, and as a result she felt she had to transfer to another post office. What should have been a lateral transfer, in fact, turned out to be a downgrade for her.

First, she's being sexually harassed, and then she transfers to another post office where she's given only 15 minutes to make up her mind to take the downgraded position, which was clearly unfair. And then they downgrade her with a loss of income, all because she was being sexually harassed and she filed a grievance because of it. Now, in the new job, she again experienced sexual harassment. Her vulnerability to this abuse causes a sense of helplessness because of the abuse of power by these male supervisors. That's what sexual abuse really is; it's an abuse of power by somebody who has power over you and you are helpless. I think, in her case, there was also a great deal of rage that she could not do anything about it, and no one seemed too interested in doing anything about the abuse.

She was transferred to a postal facility where she was working with machinery. When she started to work with the machinery, she didn't realize what effect it was having on her. She really started getting the crazies, and at times she would take something or other to help her cope with this. The interesting thing was when she was transferred away from that job to some place where there was no machinery whatsoever, she was functioning perfectly well. So now, here she is not working with machinery, doing okay for two years no problems, and then that facility closes down. They put her again in a facility where she's working with machinery, she again experiences a breakdown. Now maybe there are many postal employees who can work around machinery and don't have a breakdown. In her case she was supersensitive to the effect of machinery noise, and it was almost like a clear cause and effect relationship. She's working around machinery, she has a breakdown. They transfer her some place where there's no machinery, she's functioning. She goes back to where there's machinery, she has a breakdown, again.

In another job related-incident, she goes into the cafeteria on postal premises to buy her usual bagel and coffee, and after eating half the bagel notices what looked like mouse droppings on the bagel. She freaks out and gets violently ill. Several fellow co-workers actually saw what looked like mouse droppings. She could not return to work after that. Unfortunately, she brought that sample of bagel to a lawyer who, for whatever reason, says he lost it, which quite frankly, seems hard to believe. But anyway, he lost her evidence. But there were eye witnesses, and she knew what mouse droppings were. She grew up in a neighborhood where there were plenty of rodents around and she knew what mouse droppings looked like.

There was absolutely no sympathy from management whatsoever for this woman. I mean, it's like a very hard hearted attitude toward this poor woman who freaked out. Now maybe some people can eat something like that, and go on eating it or throw it out and keep working. But here, you're dealing with a very sensitive woman.

Now I want to conclude about her, and this is the really sad thing. Management kept downgrading, and downgrading and downgrading her to a point where she's making a decent living, you know, for somebody with only a high school education, but it's very demeaning. The way she put it, she had such high hopes going to work for the post office and using her computer skills and working her way up. Instead, she's worked her way down to the point where she's cleaning toilets, which is not why she joined the post office. She's being well paid for cleaning toilets, but it's humiliating for her to be doing this; this is not what she wanted to do. And the thing about her is that she was offered, several times, to go out on permanent disability. Here's a woman who didn't want to go on any form of welfare; she wanted to work. She wanted the opportunity to make a good living for her two children. She not a woman who just wanted to sit around and get money for doing nothing, but she wanted to be treated fairly.

She has been with the postal service for ten years, and within that time she was hospitalized thirteen times for emotional problems, all job-related. Her initial downfall was the sexual harassment by her supervisor, and then the transfer to another facility. Once she started grieving her complaints against her supervisors she said she knew she had been listed. She's convinced now that that's their goal, to get rid of her. She could stay there on the lowest rung of the ladder, lower than what she was when the sexual harassment began, but with no future. She's terribly depressed, and very angry that this happened to her. But she's not about to quit or be forced out, because she wants to provide for her family. She doesn't want to live on disability. She's willing to work hard. She's just in a rage that this is happening to her, and she has a sense of helplessness. You can almost translate that into what happens when you're sexually abused, and you have somebody who has more power over you.

An example of the eggshell theory is where somebody uses their power over you, whether sexually or otherwise, and you just feel kind of a helpless impotent rage and you feel like there's really nothing that you can do. You appeal, you go to hearings of one kind or another, and all the way up and down the line they just back each other up. And you feel like you just don't stand a chance.

OPPOSITION TO SEXUAL ADVANCES
LEADS TO TERMINATION

"This guy was envious of him ... these were women that he was attracted to himself but who seemed to be more attracted to the craft employee than to him ... he's like a little guy taking advantage of his position ... his power ... he could do this to this tall big, beautiful guy ..." Dr. Wright

This is a young, married man working for the post office. He had an excellent work record and no problems with supervisors, until he started working for a particular male supervisor. The supervisor apparently was jealous of this particular employee's alleged sexual conquests with women at the post office. He started to hassle him and make all kinds of derogatory sexual statements to him, which of course had nothing to do with his work. He was asking him personal questions about his sexual prowess and so on, which again had nothing to do with his work. It was abuse of power. This man had no business discussing any of that with this man and the employee tried to cope with it as best he could.

I believe that the supervisor probably felt sexually inadequate compared to this employee because the supervisor is much shorter. He's like a little guy taking advantage of his position, you know, his power, that he could do this to this tall, big, beautiful man. So he's going to break this man down. According to my patient, and apparently this was verified through eyewitness reports, at one point the supervisor threatened to kick his butt. This threat was apparently made in the presence of other employees. The supervisor did not stop because on another occasion he threatened the employee, saying, "I know which way you go home," or words to that effect. Other employees heard this threat and walked with the employee to his car just in case the supervisor attempted to make good on his threat. But then, the supervisor accused the employee of threatening *him* and succeeded in getting the employee fired from his job at the post office. This is retaliation.

By the way, the employee handled the situation the best way that he could, and with great dignity. He could have wiped up the floor with this guy, but he didn't; he could have met him in a dark alley, but he didn't. He's not a violent person; he just wanted to do his job.

And as a matter of fact, the vendetta against the employee by the supervisor continued after the employee was fired from the post office. The employee was initially refused unemployment benefits based on false statements made by the supervisor, and other management personnel at the post office. However, after a hearing, the hearing officer concluded the employee was not terminated for just cause, and he was able to collect unemployment for almost a year, which helped, but did not compensate him adequately for loss of income from his post office job. He's still fighting for his workman's comp — in other words there was no end to the supervisor's continuing and ongoing harassment. When he came to me, the employee was suffering from depression. And all too often it is the case.

The depression can be caused by many, many factors. It could be also loss of income, it could be loss of a position, but also, it could be as a result of underlying rage and anger which you can't adequately express; and where you have a sense of helplessness, there's nothing you can do about it. In this employee's case, it was this little manager threatening to kick his butt, knowing he can get away with it because he's boss. So the employee had to be careful. But he was just building up this rage, and this anger, which I believe, he was containing as best he could, amazingly so, because there was a tremendous amount of provocation. In a sense, he was turning his anger against himself rather than to externalize it and direct it at anyone else, and that was my understanding of the source of his depression. He's not the kind of a man who would just go in there with a gun and start shooting people. Also, there was the loss of his job, his income and his wife who left him with their newborn baby.

The sexual harassment by this supervisor and his power to have this man terminated from his postal job after ten years of service, was the direct result of his emotional and financial stress. Here's the thing: this patient, who worked for the post office for twelve years, doesn't have a history of problems with supervisors until he met this one guy. That seems to be the case with a number of these people. They don't have a problem getting along with people, or with supervisors, but then they meet one supervisor who's a supervisor from hell, who happens to choose a particular employee to hassle for whatever reason. This is a good example of this employee and his supervisor. This guy was envious of my patient, you know. These were women that the supervisor was attracted to himself but who seemed to be more attracted to my patient than to him. So he's playing with his f------ head, to put it in plain English, and was able to get away with it because he was protected, and nobody did anything about him. They just protected him, the other managers. They just lied for him.

The little manager made like the employee was a bitter, angry, paranoid, disgruntled, trouble-making employee, who's justifiably fired from his job at the post office. Nothing can be further from the truth. This man was a very stable, well functioning individual who got along with people very well. He was well liked. And you know the funny thing is that this employee was not sleeping with any of the women the supervisor thought he was. He just had female friends, and the little supervisor was jealous.

But you know, you can make somebody, seemingly, into a mean kind of a person by the way you treat them. I mean, it seems like it's possible for a manager at the post office to just start harassing somebody in a very sadistic way, and nothing is done about these people; they're still there. The post office seems to back these people up, no questions asked. You really wonder if anything is ever done about any of these people.

These workers are not suffering from physical injuries but rather from psychological injuries which are clearly a result of the behavior of their employers. You can clearly see the causal relationship. You can see it so clearly where the workers have a history of good work records until they come up against somebody, and then suddenly, for some reason, they no longer have a good work record after they get into sh-- with a particular supervisor. Then things start to fall apart for these people who just are going to the workplace to make a living. But they get involved with certain managers who want to use their power, or abuse their power, in ways that have nothing to do with their supervisory responsibilities. It's a personal thing with them. And you know, it seems like nothing is ever done about these people. There doesn't seem to be any system for getting rid of these people who abuse their power. They are still working there. When the employee is drummed out of the agency, because emotionally they can no longer work, the supervisor will find somebody else to hassle somewhere down the line. It's not like it's going to stop.

There should be some kind of system in place for really investigating the abusive behavior of management, and maybe remove them if necessary. But as far as I can tell, at least from the people I've seen, that doesn't seem to be the case. The people that are removed are the victims, not the abusers.

CHAPTER V

"All of a sudden I felt somebody push me. Before I could turn around a man said, "Don't turn around ... I have a gun ... I tried to get out of his grasp ... I could smell the liquor on his breath ... he pulled down my pants and my underwear ...I started to scream ... an official called me the next day and told me that she thought it best that I return to work the next day ... it was important for my employees to see that I was all right ... to return to work as though nothing had happened ... "   Caitlin

I worked for the postal service for twenty-five years. I started when I was twenty-six years old in a small office in the south that didn't have any delivery. In that kind of atmosphere you have to learn everything about running the office, which I did. After a little more than a year I decided that I wanted to broaden my background and learn about delivery so I transferred to an office that did have delivery, and I learned the delivery end of the business. I was promoted to a floating supervisor position, which is a supervisor who works in a different office everyday covering for the permanent supervisor who is off that day.

    My postal experience was never ever easy. In the early years the comments that I heard from all the managers were, "Why don't you get married, you're taking a job away from a veteran." There were comments after comments after comments, and the more they could embarrass me, the more they tried to. But I decided that I was not going to let that deter me from my objective which was to support my children. After I made supervisor I worked a different office everyday. I would work in the delivery section in some offices, and the mail processing sections in other offices. This gave me a background that led to me filling in when there was a shortage, like a supervisor taking a vacation, sick leave, or when an office would be left unsupervised I was put into that office temporarily. I worked in at least ten to fifteen different offices in the south in my postal career supervising because of the knowledge I had. While I was a floating supervisor, a permanent supervisor position came open at one of the offices, I put in for it and I got the job. I remained supervisor there until the time of my injury.

I was the first supervisor on duty. I would always go into the front door in order to check the cleanliness of the corridors and lobby, to make sure everything was in order for that day's activities. The rest of the employees would come in through the employees entrance in the back. It had become too dangerous for the employees to open up by themselves in the morning. Things had been happening over and over again, and I was getting apprehensive about the situation. I would inform my postmaster that anytime something happened I was going to have a sheriff report made out so that if in the future something did happen, they couldn't say to me, "Well, we never knew that it was dangerous here" or "You should have let someone know." So that's what I did. When people in the area thought everybody was out of the office, bottles would be thrown through the windows. One of my employees who opened the gates in the morning said that people would throw dirt at him, curse him and things like that. So we started having two people opening up the office, and then I would come in.

The post office was right across the street from an area where a lot of drug deals went on. The sheriff had a habit of sitting in a van in the post office parking lot doing surveillance. They had a television monitor in the truck so that whatever happened they caught it on camera and taped it. I didn't usually go in the back because I would park my car in front so that the employees could park in the fenced in area where it was safer, and they would not be exposed to the public. There was not enough room for everyone to park in the back. We owned a piece of property next to the postal service so I asked the proper authorities if they would enlarge the parking lot, and put a chain link fence around the whole thing so that the trailers and trucks could park back there. They didn't want to do it so I was parking in front so that there would be room for the employees to park in the back.

This one morning when I reported to work at 5:30 a.m. the first thing I saw was that sheriff van sitting in the parking lot. So with that I never even considered that anything could be wrong. I went and unlocked the door and I got inside, but there was still another unlocked door that I had to open up. All of a sudden I felt somebody push me. Before I could turn around a man said to me, "Don't turn around and look at me, I have a gun. If you turn around I'll shoot you." He grabbed me and immobilized me. I couldn't think. I guess I was in such shock. I tried to struggle with him. I tried to get out of his grasp, and he kept saying to me, "If you don't stop I'm going to shoot you." It took everything I could muster to make myself stop struggling. My struggling was an involuntary reaction. I was trying to get away from him. Finally I said to myself, "If you don't stop, he's going to shoot you." So I finally got myself together enough to stop struggling.

It's not really clear how everything took place. Somehow I was thrown against the wall, and somehow I ended up on the floor. I had taken a course on rape at the sheriff department, and they said, "Try to talk calmly to the attacker. Try to reason with him." I was so, I don't know how to say it, crazed. I was a wreck. My mind was running in a hundred different ways. I couldn't get myself to ... I couldn't talk. All I kept thinking was, "Maybe he'll shoot me in a place where I won't die, the shoulder, the leg." I wasn't paying attention to what he was telling me to do because I was trying so hard to calm myself. I was thinking, "If I could just calm myself down enough where I could reason with him." I was so involved with what I was thinking that I wasn't paying attention to him, and he started strangling me.

I finally pulled myself together enough not to scream, because they told you to talk low to him. So I finally pulled myself together enough to talk low, and try to reason with him, which is what I did. I told him that he could do whatever he wanted as long as he didn't hurt me. I told him there was no one else in the building. But there were two other employees in the building; they were unloading the truck in the back. I could hear the music through the door but, if you did not know they were in there you would never realize it. Only because I knew they were in there did I hear the post cons[1] being moved around. And I was afraid that if I told him they were in there they would be in danger too, since he had a gun. So I figured, the best thing was to try to get myself out of it without involving anybody else.

I wanted him to stop strangling me so I said, "Please, just don't hurt me. Whatever you want. Fine." I noticed that when he first grabbed me he must have been drinking all night, I could smell liquor on his breath. That smell that people have when they drink for hours and hours. He was leaning on his left hand and the bottom part of his left arm. I noticed that he had a wedding band on. I said to him, "Think about what your doing, think about your wife, think about your family, your mother, your father. How would they feel if ... I don't know what I said to him, I just said anything that came to my mind.

When he first got me on the floor, he pulled down my pants and my underwear. When he finished with me, he said to me, "You're married, aren't you?" I wasn't married but I said yes, I was. I thought that maybe he would feel sorry for me if I said that I was married. The only thing I remember him saying is, "I'm going to leave now. Don't turn around and look or I will shoot you." And I waited. I laid on the floor with my face turned around. I was afraid he was standing there watching me. I waited and I waited. And then when I thought he was gone I quickly got up; I didn't even pull my pants up ... I just ran and unlocked the other door. Of course, I'm tripping over my pants as I'm trying to get to the other door. I got in and locked it behind me and I just started to scream. I started to scream for the employee that was working. He came running over to me, and of course, I was incoherent, and I kept saying, "Just call the sheriff, just call the sheriff."

---

[1] Metal shelves on wheels that hold mail.

He called the sheriff's office and told them that something had happened. The deputy wanted me to get on the phone, I did and I told him what happened. The deputy wanted me to stay on the line until the sheriff got there. When the sheriff and several deputies arrived, I told them what happened. They asked me questions, and as the sheriff was taking my statement, the phone rang. I picked up the phone and said, "Post Office," all I know was there was a hang up. There was a female deputy there and she looked at me, and then looked at the sheriff and she said, "I'm going to take a run down the street because there are pay phones on the street down there." The sheriff asked me, "How long has that van been there," I thought the van in the parking lot was a sheriffs van." I told him, "It was here when I came to work this morning." The sheriff said to a deputy, "Find out whose van that is." Until this day, I still think it was their van, but I don't know where the deputies were that was supposed to be in it. I called my postmaster and told him to come in because I had been attacked. He and another supervisor came in and took me to the sheriffs department.

What happened after the rape was just as horrendous as what happened during the rape. I think they treat dogs better than they treat women. I can't believe the disregard for human feelings.

I got to the sheriffs department and there was no one there to interview me, so I had to wait. I think I waited about forty-five minutes for somebody to come in. I think they had to call somebody from home. The deputy that interviewed me seemed annoyed with me because he was awaken and had to come in to interview me. I was in no mood to be cordial at this point. During the questioning he said to me, "Well, did you see the gun," I said, "No, he told me not to turn around." "Well how do you know he really had a gun." "I wasn't taking any chances. He said he had a gun, and as far as I was concerned, he had a gun." While I was waiting for this deputy to get in I had written down on a piece of paper as much as I could remember because I was so afraid, and so upset that I would forget.

I just told him the description: "He was husky, white, and freckled, and all this I got from his left hand. And he was blond." "How do you know, did you see him?" I said, "No, I didn't see him." "Well how do you know he was blond?" "That was the color of the hair on his hand." I said, "Are you going to look for him?" "No." "Why?" "Because you can't identify him." I said, "I can identify his hand." "That's not good enough." He said, "We're not going to waste our time looking for anyone when they can't be described." At this point my family had come to take me home.

The postal inspectors and the postmaster had come to my home and the postal inspector wanted me to go to the hospital for a rape kit. They drove me to the hospital. When I got to the emergency room there was no gynecologist available. I waited from about eight in the morning until 2:30 in the afternoon for a gynecologist to become available. I was on a table with a sheet over me in a room all by myself. Finally, a doctor did the examination and took pubic hairs, and hairs from my head in case they can match them to the rapist. Why they wanted to bother me since they weren't going to look for anybody, I don't know. But he didn't cut the hairs, he pulled them out, one by one from my head, from my pubic hairs, one by one. I can't tell you about the excruciating pain I went through as he pulled out at least twenty hairs from each area. I don't remember much more after that; I don't even remember going home.

This was a Sunday that the rape happened. The next day an official called me and said that management thought it best that I return to the office the next day; that it was important for my employees to see that I was all right; I was told that I was to return to work as though nothing had happened. I made an appointment with my medical doctor just to check me out and the next day, I went back to work as if nothing had happened. Emotionally, it was too much for me. During the afternoon I called a higher official and said, "If you want me to continue working, I cannot stay here. You have to find something else for me to do." After that I was sent to another facility to work. They didn't know what to do with me so they placed me in some innocuous department.

The medical tests found that I had a urinary tract infection, which I confided to one of the female supervisors who knew I had been raped. She said to me, "Oh, you can get that from anything." I said, "I have never had a urinary tract infection in my life. I can only have gotten it from what the doctor said it was from, "Dirty fingers, dirty what-have-you, and whatever else." I worked there for as long as I could. Two weeks later I called up Human Resources and I told the manager that it was "absolutely impossible." I just could not come to work.

I found it impossible at that point to function. Instead of getting up and taking a shower, getting dressed, doing my hair and putting my makeup on, after being up for two hours I was wondering around in my home. I could not manage any time schedule. And I was totally exhausted. So an official made an appointment for me to see a psychiatrist. I kept the appointment, and he said, "I usually do evaluations for the post office if an employee suddenly starts running around on the work room floor, or screaming or doing something that's abnormal; they usually send them to me and I evaluate them." He said, "From what I was told this isn't the case with you. So why don't you tell me what happened." And that's what I did. He asked me questions; he interviewed me, and after it was over he said to me, "I'm not suppose to tell the employee the findings of my evaluations, but I'm going to tell you." He said, "You are not to go to work until you see a psychiatrist and get medication, and after you have been under that psychiatrists care, he will determine whether you will be able to go back to work." He said, "Ask my nurse if we take your insurance and if we do, I'll take care of you." I told management what the doctor said but was told, "No, I'd rather you find your own psychiatrist so that we can keep this doctor to evaluate for us."

I did not know how to go about finding a psychiatrist and I wasn't capable of doing this. My lawyer suggested I see a psychologist. The psychologist who I saw, in turn, tried to find a psychiatrist for me to go to because at the point I was at, she was afraid that I would have to be hospitalized. She was intent on getting me on some type of medication as soon as possible. So I had an interview with three different psychiatrists, and I chose one. He put me on Amitriptyline and I was out of work for about a year and a half.

MY RETURN TO WORK

At some point I felt that if I went back to work I would progress. I felt that maybe staying home was the wrong thing to do. I still had problems with my timing, but I felt that if I had a goal, a focus, that I had to be somewhere at a certain time, I could get myself back on schedule. So I went back to work. I was not well-groomed, but I went back to work. I was still incapable of taking care of my personal habits. I was clean, but I did not have make-up on, my hair wasn't done up. You could see that I was still a little disheveled; I still wasn't together, but in time I felt that I would counter that. And in time I felt that I had gotten better and better. I felt that I was useful again, and I was getting myself back to where I had been. When I went back I started working two hours a day, then I went up to four hours a day. I eventually worked my way up to eight hours a day.

At first they didn't know what to do with me so one of the managers took me into her department where I stuffed envelopes and did odd jobs. Then someone who worked in another department that I had known before I was injured, and who knew that my work was good, asked management if I could work in their department, which I did. She knew of my professional reputation. She knew that I was knowledgeable, and responsible enough that when there was a problem in an office they could put me there and I could keep the office running without a problem. Besides that, I was on the Board of the Woman's

Programs. When we did the block facing for the Zip Plus Four, I was on a team that did that. I block faced out every street by different towns. So I had a lot of responsibility as a supervisor. It wasn't just that I ran one office. A lot of postmasters had asked me if I would come to their office when a supervisor position was open. I felt that I had a good professional reputation. I had been to the Management Academy for three weeks just for the Women's Program, and also took advantage of other courses that were offered.

When I got into the department I straightened out this manager's files, I learned all the computer tasks, I did all the reports, I handled everything so that she could do her work. When she wasn't there I had enough knowledge so that I was able to, and did everything. While I was in that office, six different people trained for that job, and I trained them all. At this time I was working four hours a day. A lot of times when the manager was away somebody else would take over the job temporarily. I wanted to take it over, but they said to me, "As soon as you are working eight hours you could have that detail." So I kept building myself up, striving toward that. I learned everything that I could. I kept it going while the manager wasn't there, and when I was back to eight hours a day they told me they couldn't let me do the detail because I was an injured employee. Meanwhile, I knew of other injured employees that had taken detail. I knew that what they were telling me wasn't true.

I had gone back to work in good faith. I was working very hard for them. Whenever they needed anything done, "Caitlin, you want to be the Special Projects Coordinator?" which was a job all by itself, but I did it along with my regular job. I was the primary coordinator, I did everything for the Special Projects Program. It took me hours, but I did it.

Other departments would come to me and ask me to help them out, and I would help them. Postmasters would call me up with questions, I would help them. And they knew of my capabilities, they knew they could count on me. And I knew they knew this because even after I left work for that year and a half, I was speaking to one of the girls I worked with and she said, "One of the postmasters called and asked for you and I said you weren't here anymore." I said, "What did he want?" Well, I asked him if I could help and he told me that it wasn't about this department, but every time he needed a question answered he would ask 'Caitlin,' and you would help him." So my former co-workers knew that I helped whenever I could. I certainly did not know all of the answers, but if I didn't know, I knew who would know, who they could ask and who could help them. I became very entrenched in the job.

I began to become very frustrated when the details were going to everyone else, and I was training them. But I was not good enough to have the details. This particular manager was away from her job for a year and a half while acting as a manager in another department, and in that year and a half I trained person after person, yet, I never got the job. When the manager wasn't there and there was no one else, I did the job. I did the job at a level 15 pay, and they told me they couldn't give me the money, or the title, but I could do the job. It just got more frustrating.

THE THREATS AND THE ABUSE

I always felt that the person who attacked me knew me. I don't know if it was another postal employee or what. I can't say who it was, but I always felt that I was never safe. I was afraid to walk in and out of the building at work, and I told this to my psychologist. I told her that I didn't want to appear paranoid. She said, "Why don't you just make sure that when you walk in and out there's somebody else there for you to walk with." Which is what I did. So I had mentioned to one of the employees that I didn't want to walk out alone, and asked if they would walk with me. From that time on, I walked in and out of the building with other people. I wasn't by myself.

I came into the office one day and turned on the answering machine, and there was a voice on the answering machine. I don't remember all the words that he said to me, except, he said, "I'm going to kill you." I was stunned. I thought that he had nerve to leave it on the answering machine at work. I had been getting hang-ups at home, and at work but I didn't think anything of it until I heard the answering machine and I got that message. I went to the manager and asked him to come into my office and I played the tape for him. He said, "Oh, that's just somebody playing a joke on you." And I said, "I don't know. I don't think so." He said again, "Oh no, that's just someone playing a joke on you. I know who said it." And he pushed the button and erased the tape. I was horrified. I said, "You erased the tape." "No I didn't." "Yes you did, you erased the tape." Nothing was ever done about it.

About a week later, I played the answering machine and there's another message. On this tape there was heavy breathing, and sighing, like someone was doing something. This time I took the tape out of the machine, and I went to one of the managers and I said, "This is two messages now. If anything happens to me, the first thing they are going to say is 'why didn't you inform us' so I'm reporting this to you, and I'm notifying you that I'm sending this tape to the postal inspectors," which I did. I never heard one single word from them. I made it known around the office that I had tapes that I had sent to the Inspectors. After that, I never got another phone call at work.

But still I was getting the hang-ups at home. My psychologist said to get an answering machine at home and use that. But before I got the answering machine, I answered the phone this one day and the voice on the phone threatened me and made lewd remarks to me. I said myself, 'maybe you're being paranoid' maybe it isn't for me. I have daughters. So I said to him, in the best motherly voice I could muster up, "Do you know who you're speaking to." And in the most low, chilling voice you ever heard, he said, "Yes, Caitlin. I know its you." I was frantic. I burst into tears. I dropped the phone. I was crazed. I called the phone company to get them to try and trace the call but all they told me was they were not allowed to give out that information. He's allowed to do this to me over the phone, and they are protecting him. I don't understand a law like that. I had not grasped the sense of a law like that. Anybody can use that instrument anyway that they want. So naturally, I had no other recourse, that's when I got the answering machine.

Between the agency lying to me about getting the job and these calls, I was crazed. I had gone to personnel and asked to see my file. The girl said, "Well, I don't know." I said, "Why not, it's my file. I know what happened to me. It's my file, it's my personal information." And she said all right. In the file I found a letter from the postmaster of the post office where the rape took place. The notes said: "No one in the post office heard anything, saw anything and no one could verify anything happened that morning." There was more than that in the file, but that was the gist of it. They were talking about the rape, but there were two employees there. They were all the way in the back of the building unloading the truck out on the deck. It would be impossible for them to hear anything from the back. I burst into tears. I was frantic. What upset me was that they were trying to say that the rape never occurred. They're writing down that "We don't know that anything happened, nobody saw or heard anything." This is abuse. I went through all this torture and now you want me to read a piece of paper that you don't think it happened. I ran out of the office. I went home, I was sick, and I ended up in bed. I was a mess.

The manager I worked for was so afraid of other people who might be more knowledgeable then her. When another employee came to work with me, the manager was so secretive. I said, "Look at it this way, you can learn something. Don't think of it as them knowing more than you, pick their brain. Take the knowledge they have and use it for yourself. But she couldn't understand that concept. But she liked to ridicule me so that I could look bad in the eyes of other people. When she would introduce me to other people she would say things that would make me look unprofessional, or unknowledgeable or she would just ridicule me with, "Oh, this is Caitlin, she's my assistant. She's very good, but you should see her when she walks into walls," things like that. I finally said to her, "Please don't say things like that." "Why, I'm only kidding." "But suppose I put in for a job that one of these people were choosing a manager for, and they say 'Oh, not her, her supervisor said that she walks into walls." I said, "You know, you are making me look unprofessional, and it looks bad for my professional reputation." But she couldn't understand that, or didn't want to understand that, and she did this purposely. She had her own agenda.

She would constantly say that my work wasn't done, but yet, all the reports were done every month. All the paper work was in. I had to send the reports by computer to Washington and that was even done. If my work wasn't done, Washington would have called to ask why isn't it done. In meetings with upper management, she would constantly talk about me in a demeaning manner. The only reason I knew about this is because people who were there would say to me, "Gee Caitlin, your boss said this about you," or "She said that about you," and, "I didn't know you were not well." And I would say, "What did she say about me," and they would tell me. It was just demeaning and mean spirited things about my personality, or the way I dressed. At the time she told people that I walked into walls I was taking 250 mg of Amatriptoline a day; I was lucky I got to work in one piece.

I was only taking that medication because as far as I'm concerned, upper management didn't want to protect its employees by make their premises safe, and I was a casualty of that. I was raped on their property. So instead of making fun of me, they should have been giving me accolades, and saying, "Thank you." Something. Some show of gratitude. Some form of something, but no. Like I said, when I went back to work, I did it in good faith, thinking, "I'll do whatever I can for you," meaning the agency. But I expected in return for something to be given back to me. Before the accident, no matter what department I worked in, no matter where I worked, people would stop in my office and say, "If you ever want to work for me please come and see me," because they saw the quality of work that I did. They saw how I worked. And that's not something that *I'm* saying, it's something that *everybody* else saw.

Getting back to the psychiatrist that the agency first sent me to, the reason the agency wouldn't let him take me on as a patient was because he was so adamant about how bad I was that he told me what his diagnosis was going to be. Ordinarily, under no circumstance will a psychiatrist tell you that. It's suppose to be private between them and the employer. I think because he was so concerned about me, the agency did not want him to take care of me. But, they did not try to help me find another psychiatrist. They did not assist me. They did not send me to the Employees Assistance Program (EAP), or anything like that. They did nothing for me. They just left me out there to flounder.

As I say, when I went back to work in good faith I was a manager. I was part of the management team. I did as they directed. I took care of their finances, I made sure they made a profit, I made sure that the offices ran smoothly, I went anywhere they sent me. Being part of the management team I *expected* them, in turn, to treat me in a fairly humane way. I was part of *them*, but instead they treated me like I was nothing to them, like I did nothing for them, that I didn't matter. I have seen them treat craft employees like that. But I thought that they would take care of me, that they would help me. But the way they treated me, I felt that I was being abused on top of being raped. I expected them to tell their doctor that they would like him "to stay as an independent and help her find another psychiatrist." I expected them to treat me with some dignity, that would measure the position I held for so long in the company.

Like the tape that the manager heard. He heard the first tape, he heard the words "kill" and yet, he denied knowing anything about the tape. He claims that it never existed. This is another manager, this is *my* manager. Part of a manager's responsibility is to see to the well being of their employees. And as far as I'm concerned, he was not concerned about my well being at all. He actually lied, barefaced, saying there was no tape when he did hear the tape; and second of all, somewhere in the postal inspector's office there is a second tape. I didn't expect him to lie, but he did. They blatantly lie about what goes on.

All of these things were going on, and I was becoming more and more disheartened because the promises were never kept, lies were being told and it hurt me. I went back to work and did the best job I could for them, which was a good job, so everybody said. I even got a $250 reward *because* I did such a good job and I received another award from Washington *because* of the good job I did. But everyone said that because I did such a good job, my boss wanted to get me out of that particular department because I made her look bad. There was another employee who went on detail to Washington. While he was gone, my boss told me that there wasn't enough work in the office for me to stay there. So they had to find somewhere else to put me. They put me into another section, and when this other employee returned I said, "They told me there wasn't enough work in here for me to stay." "That's not true Caitlin. There's more than enough work." I had gone back into the office several weeks later and I was told that it seems that the way they decided to do the reports, that I was doing alone, was to divide the work among themselves. Also, another report which I did, my former manager decided that when I left the office the reports were too cumbersome to do and she stopped the report completely because she, the manager, wouldn't do them.

So it was things like that, abuse after abuse, that disheartened me and made me dysfunctional. It really did. I *had* been progressing, I *was* getting better. I worked myself back up to eight hours a day because I had a goal in mind. I wanted to get my career going again. But I finally realized that they were never going to move me. They had decided that they were just going to put me any place they could to keep me out of the way.

When I returned to work, I did excellent work. I got a letter of commendation. When I was out sick for my operation, one of the managers had asked the staff to do SOP's on the computer. One of the business reps said, "Wait till Caitlin comes back, she has the best notes of all of us, and she can type them up." So here I was in this particular section for a very short time, and already I had the best notes, and everybody else has been there for two years.

Within a year of returning to work, I started to regress because of the broken promises and lies. It had nothing to do with not getting a job. It had to do with, "My word is my bond." And I expect that if you say this is how its going to be, then this is how it's going to be, unless you say, 'Caitlin, this came up and because of that we have to do this,' then that's fine. Don't tell me one thing, and then three months later say "I never said that" or "That's not the way it happened." So I had vacation time, which I took. When I returned to work, they put me in a cubicle in the back of this particular section and had me read books. As long as I worked in that facility, people always had access to the ladies' room. If you had to go to the ladies room, you would leave your station and you would go to the ladies room. Not that you had to be watched. But, I was told by the manager that I was not to leave my cubicle unless I asked permission. I was told that I was not to leave the section, I asked, "Why is that?" "Well, every time we look for you, you are not there." I said, "Well maybe I'm either in the ladies' room or when I have to do envelopes I have to go to AIS to get them approved, then I come back. I have seven for each customer, and it takes time to do some of the things I have been assigned to do. If you want me to sit here and not move I won't be able to do those tasks. I must have them verified." I just felt that this was one more way of abusing me when every one else in there comes and goes as they please. I have been a supervisor for fourteen years. I have the most knowledge in that place. Anybody will tell you, I am a diligent worker. I never had to be watched, I was always working. And if they had to do that to me it was like a knife in the back, that was cutting.

  I was in an accident and I had to stay off my feet for awhile. Between vacation and sick leave, I had at least 12 weeks of leave and I decided to take that leave. While I was out, a position became available. I put in for the job, and called up the manager of Human Resources and told him that I was afraid if I don't come back to work before they chose someone for the job that I won't get it because I'm not there. I was told, "Caitlin, they can't do that because of the American's Disabled Act. Even though you can't work right now, if you are the person chosen for the job they have to hold the job for you. Somebody else will do it until you get back." I said, "Oh, all right." So I was planning to go back to work.

  During this time I was seeing my psychiatrist, but by that time I had regressed, and I said to my psychiatrist, "I cannot deal with work anymore. I can't get myself together, I can't focus, I cannot concentrate." He said to me, "I want you to try. I know you are having problems, but just try and see how you do." I said all right. But the Amitriptyline had stopped working and I was a mess, so he took me off that and put me on Prozac. He said that I could not work until further notice from his office.

  I had the interview for the job over the phone, but then, I had not heard from them for some time. I was speaking to someone at work, and they told me someone else had gotten the job. I said, "Really." "You mean to tell me they never called and told you." I said, "No." "Well so-and-so was suppose to call and tell you that someone else had gotten the job." "He never called me." At that point I was so sick, and I was on Prozac now, and it didn't even matter.

My supervisor sent me a letter stating that the agency doctor said that there was nothing wrong with me and if I did not get back to work she was going to put me down for AWOL. I called the manager and said, "Don't you have a letter from my doctor that says I can't come back to work." "Oh, Oh, yes, yes, I have that letter." "So why are you sending me this letter from the agency doctor." "Well, you're capable of coming back to work." I said, "As long as my psychiatrist says I can't, then I can't. And I'm not capable of coming back to work right now."

She then sent me a second letter saying that the agency doctor says that I can come back to work. When I called her she said that I have to have my psychiatrist fax an updated medical report. So he sent his report to the agency injury compensation office, and I told my supervisor this. She called me back and said, "Injury compensation sent the report back because they had not asked for it." I asked her why a medical report would be returned even if it were not asked for. I said, "Wouldn't you want to keep it in the file just so you would have an updated medical report?" She said, "I have to have another one." My psychiatrist said that he faxed her one. She sent me a third letter stating, again, "The agency doctor said that you can come back to work, and if you don't come back to work I'm going to fire you." I called her up and said, "My psychiatrist faxed you that letter." "Well, I never received it." When I said I will have him send it out again, that's when they fired me. Now that is abuse, upon abuse upon abuse. That is ludicrous that anybody should be allowed by the other management levels to treat their employees like that. With that, I spoke with my psychiatrist and he told me that if I go back to that atmosphere, between the co-workers and the way management treats their employees, he said that I will never get better. He said the best thing for me to do was to try and retire. He told me, "You have to face the fact that you will never be able to go back there to work."

I am giving this interview because, through the years as a post office supervisor I have seen the way management treats the craft employees, and the way management talks about them. I have always tried to treat my employees in a more sensitive manner to protect their well-being, as well as my own. One of my employees used to address me as, "Good morning, the human supervisor," because I refused to play the games that other management played with the employees. I refused to degrade them like they did. I just feel that no one should be allowed to treat their employees like that, and the post office seems to get away with it.

When you are a craft employee you don't always see the real abuses that management does to its employees. Being in the management group I was able, with my own eyes, to see some of the things that they do and wonder how they get away with it. Doesn't Washington say to itself, "Gee, there are an awful lot of EEO complaints and workers' compensation claims in that District, what's going on there?" Is the abuse all the way up in Washington? Is the abuse that ingrained in the postal service that they don't even question why so many things are going on?

Maybe with this book, people will begin realizing how these employees are treated and maybe things will be turned around. Someone will have to be called to task for the things that have been done to federal employees.

CHAPTER VI

"I was attacked on the job by a postal customer; he beat m. He kicked me in the stomach, but I lived; ended up in the hospital with blood in my stomach ... had to have my stomach pumped to get the blood out, and they did absolutely nothing ..." *Valerie*

I've been dealing with excruciating pain in my body for years now, and going from doctor to doctor has never brought any relief. I've taken a lot of medications that subside the pain a little bit, but never takes it away completely. In the beginning, it was hard to deal with certain things like my children and my husband; I just didn't want to be around them any longer. A lot of times I blamed myself because I felt like if I hadn't gotten hurt, they would have the extra income, and things wouldn't be so tight, and my husband wouldn't have had to leave to go to the military to try to supply me with the extra insurance to make sure that we'll always have some type of income.

LIFE BEFORE THE POST OFFICE

Before I went to work at a post office on the east coast, I had a job with a school board which I liked a lot because it dealt with disabled children. I really liked that because they really depended on me, and no matter what you looked like, they didn't see the color; they just saw somebody was caring about them. That job meant a lot to me, but when the post office offered me the job, I went. Prior to that I was always very active, like I participated in Walk-A-Thons every year. Being active was a part of me.

I grew up in the south. I got a scholarship to college and I played ball. I know a lot of people called me a "tomboy" even after I had kids and everything, but that was just part of me. And fishing, oh God, I loved that stuff. Me and the kids and my husband used to go to carnivals, and we'd take the kids out of town and go on cruises and stuff like that, and it was just great. I mean, we went everywhere together, we did everything, it was so nice, you know. We didn't think about money or what tomorrow's going to bring. I was healthy then. I had my tonsils taken out when I was younger, and once I had a broken limb, but I got that broken limb playing ball, so it didn't matter. As a little kid I had asthma, but I didn't let it bother me. The doctor said, "Oh, you shouldn't run and you shouldn't..." I did it anyway. I ran track. I was in the marching band in high school, I mean, I just went all out because I wanted to be, you know, I wanted to excel. When somebody said I couldn't, I was going to prove them wrong, and I always did. People would tell my mother that I was the smartest kid she had, and it was twelve of us.

I bought myself my first home when I was 20, and I had my little girl nine months later. When my daughter was about 3 months old I went and worked on the railroad. When I started, I was one of the first women to ever work on the railroad; they denied every woman a job because the work was so strenuous. On the railroad I was a switchman, and my job was to throw switches. They don't throw themselves. You get down and you throw these big solid steel switches with your hand, and you pull those switches over, and that's hard work. Not only that, I had to get up on the box cars while the engine was pulling them; everything was always moving when I got up and off those things.

The fastest I've ever gotten on one was when it was doing about 25 miles an hour, and you have to get up there, and you have to count yourself to make sure you don't miss that step because it was very dangerous; you can always step underneath the wheels of the train or the boxcar. I liked the excitement. I would get up there, and at the same time I had to pull pins, meaning I had to separate the boxcars from the engine or separate boxcars from boxcars in motion. I had to give signals to let them know, "O.K., I'm ready." I did the pin with my foot while holding onto the boxcar. I would give them a signal with my right arm, holding on with my left forearm wrapped around one of the bars, and I'd give them a signal to stop. When it just eases up a little bit, I flip that pin up with my right foot. It took a lot of power, too, because that thing is heavy because you're trying to uncouple a train; then the train will move off real fast, the boxcar soon slowed down, and then I get off, and do whatever I have to do.

I did a lot of running on rocks. It was no solid ground, always rocks and sand, and I would walk in on the rails across bridges that are in the water. Alligators and snakes would be in the water, but it was great and I was so strong and healthy back then. I had to be healthy because I worked 12-hour shifts. I would only get an 8-hour period in between there, and I went back to work again.

I stayed with the railroad 15 months and from there I went into construction. I did whatever it took legally to take care of my family. I tried to get on with the longshoremen but they wouldn't hire women back then. I went down there every morning, and they didn't want that, but I wanted the money so that I could do more for my children than I had when I was growing up. Finally, I went down to the Laborer's Union and they put me on as a laborer. I drove a stick shift, and they had these big trucks, and I would get in there, and I would drive these trucks around. The guys said, "God, that girl can drive that truck. Where'd she learn to drive a truck?" And then we would put up sheet rock; we laid blocks; we poured floors, tied steel, put roofs on, whatever it took; that's what I did. I also operated heavy equipment, bulldozers, front end loaders, bigfoot, steamroller. I was also a concrete conveyer operator, and I tied steel for the State road; so I did a lot of work. It was hard work, but I enjoyed it. I even kept my little figure back then; I was kind of petite back then, but I have to say I was real healthy, and I enjoyed myself, even though I worked hard.

I enjoyed myself and I enjoyed my daughter; my son came along later in 1983, and at that time, I was doing construction. I had to stop working construction, but then I did get a little job as a security guard. I had to go through the police training, and I was pregnant at the time, but I did my running and leaps over the wall. In the beginning of my pregnancy it was real hard because the doctor would always tell me I had to stop because I was going to lose the baby; it was too much pressure. I'd say, "Nah, it's not, I can handle it." When I was five months pregnant I decided to stop working. I wanted to sit down. I had saved some money, and I said this will do me until I go back to work. I had my own home so I was fine. My son was 8 pounds, 11 ounces. Oh, that was a handsome, little fat boy. He was about 3 months old when I applied for a job as a correctional officer with the state. I worked as a correctional officer at a women's penitentiary. I worked nights for awhile, and then I got on the 3 to 11 shift. I was healthy, and I felt good about myself because I was young, and I had achieved all these things without mamma and daddy. I raised my kids with the love of God. Welfare was out of the question.

I've had my share of tragedy. When my daughter was two, she was molested by the baby sitter's nephew. That was hard. You look at it on television, you hear people talking about it, but you never expect it to hit home. And when it hit home, I didn't think that I could take it, but I did, because more than anything in this world, I knew my little girl needed me, and I didn't have anybody else that I could lean on outside of God. I looked to God and that was the only peace that I found. My daughter had some problems growing up, she would fight the little boys. We both went to therapy and she never spoke about what happened, even to this day, and we never pushed the issue. The therapist would say that maybe in the back of her mind she's rebelling against what happened. My daughter will be 15 in May. She's saved and filled with the Holy Ghost, and she depends on God for everything. I thank God that I was an influence on my child.

Afterwards, we went on with life, and I tried hard to get back into the workplace. I took any job I could find and I was working but I decided to move east with my children. So I sold my home and did just that. I put in an application with the school board for a bus driver position. I was determined to get that job because I wanted to have the schedule that my kids had. At that time I was by myself raising my two children, and just moved to a new city. I wanted to spend more time with them because quality time is important with children. One day I just made up my mind that I was going to get that job. I prayed about it and one Sunday the lady who does the hiring and training called and said, "Are you still interested?" I said, "Yes, I am." That was in 1984. I passed the test with flying colors.

I loved that job because I loved working with the kids. Those kids needed me; they were Special Education children. They were not just kids I transported back and forth to school; they were like my own children. They were big kids, but they still wore pampers because they still messed themselves, and it was part of my job to change their pampers and stuff like that. And I did it because I didn't want those kids to sit there in no mess, and I didn't want them going home to their parents with that kind of stuff. I just like helping people.

THE POST OFFICE YEARS

"Working in the post office, while on limited duty ... the agency says ..."Oh, it's nothing wrong with you ..." I'm sure that it's not just with postal workers. There's other people that experience what we experience in the post office, and I think this book will let them know that they're not alone ..."   Valerie

I finally got the call from the post office. It took them several years to get around to calling me, but they finally called me. I took the driving test and the man who gave me the test said I was the first person that he had ever seen that passed the driver's test with no errors. When I finally got into training with the post office, I had to train in their vehicles, and back then they were half ton jeeps. I was offered a job as a tractor trailer driver, and why I didn't take that job, God only knows. And I mean this man begged me, he says, "You can do it, I know you can." I asked him what hours would I be driving, and he told me sometimes early in the morning, as early as 2 o'clock. I didn't want that job because it would take me away from my kids that early in the morning. I wanted to have a job where I had the hours with my children so they'll be going to school when I'm going to work and coming home when I'm coming home. I can pick them up from child care, and the school and bring them home with me. So I turned down the job as a truck driver for the post office and took a job as a letter carrier as a Part-Time Flexie (PTF) instead.

While I was on probation I had my first job-related injury. The mail bag is very heavy because of the letters and magazines and whatnot, and the bag can be awkward to carry when it's heavy like that. The magazines make that bag real heavy. Well, one day I was carrying the heavy mail sack and I fell, and when I fell, I experienced a shoulder separation. But I was afraid to tell my supervisor because I was a PTF on probation, and I saw PTS's come, and I saw them go. I said, "God, where am I going to go now?" If I tell my supervisor I injured my shoulder during my probation, I'm gone; I know I'm history. Instead of going back to the office and going to the hospital, I sat that satchel back on my arm, and like Mel Gibson in Legal Weapon, I popped that bone back in place myself. That was the most painful thing I had ever done in God knows when. My shoulder swelled up and everybody said, "God, you got to get used to that bag." At that time, I never told anybody about the shoulder separation. Anyway, I had scraped the skin was off my arm and my legs when I fell, and being this was in the summertime, I had to make sure that nobody saw the scraping. After that day, I never wore the shorts; I always wore long jeans so no one noticed the scrapings on my legs and on my arm. I went to that job everyday in so much pain, but I would continue to put that satchel on my arm. It wasn't because I wanted the job so much; it was because I had two children to support. I knew what I had to do, and that was to take care of my children no matter what. I could not depend on anyone else.

I went to the doctor on my own that same week. I saw a doctor who took X-rays, but he said the X-rays didn't show anything. He gave me massages and put me on heat and all this stuff, but the pain was still so bad. I had tingling up and down from the shoulder down to my hand. It felt like my hand was asleep all the time. My hand would swell and go down, swell and go down, and the doctor said, "You know you need to tell these people that you got hurt because you're going to lose the use of your arm." "No, I can't tell them because they would fire me. I need this job to support my children." He said, "What do you care about more, whether or not you're going to be able to use your arm or the job?" "It has nothing to do with this stupid job, it's my children." I told him, "I refuse to let my children go ragged." So I worked.

Finally I was off probation in November, but in December my doctor said, "I'm sorry, you must tell these people." I said, "I'll tell them now, because I'm off probation. I just didn't want to lose my job." In December I went to my supervisor and another official and told them about the injury and, I mean, nobody asked me if I was OK or nothing. The first thing that came out of one of the official's mouth was, "Are you on probation?" I turned and I said, "No sir." He got pissed, and he swung himself around and told the other supervisor, "Deal with her." This supervisor says, "Why didn't you tell me back then." "Don't ask me why because you know why." I said, "You know I would have been history. You see this man right here today asking me if I'm on probation because he was ready to let me walk out the door if I was on probation. So that's why I waited." He said, "That's why your arm was so swollen on the shoulder?" I said, "Yes." He said, "Everybody thought it was because you were carrying the satchel; why didn't you go to the doctor?" I said, "I did." "But you never told us." "I went to the doctor on my own, and I've been going to the doctor ever since it happened."

In December my primary doctor referred me to a chiropractor because I started having a lot of pain in my spine and head. His diagnosis was a lumbar problem. I can't remember exactly what. He started adjusting and pulling my neck and cracking my head and cracking my back and it felt good only for a little while. But the pain was always there. I would have to sit up and put my elbows on my thighs and push my hand behind my ears and push my neck up myself with pressure to try to release the pressure off my spine. I got so used to that because that felt like the only thing that would release that pressure, but the tingling and the numbness never stopped. The pain never stopped, so I tried going to another doctor. I mean I just went around trying to find somebody that could help me, that could tell me why am I hurting like this, why do I have this tingling pain like this. Within six months time I went back and forth to at least at least 15 doctors.

JOB-RELATED INJURY NUMBER TWO

"I was attacked on the job by a doctor, a postal customer ... he beat me and kicked me in my stomach ... I ended up in the hospital with tubes in my stomach to pump out the blood ...The supervisor called me the next day. He wanted to know when I was coming back to work. I hung up on him."   **Valerie**

I was delivering mail to this particular customer who felt like he should have gotten his mail delivered at 8 o'clock in the morning, but I'm the carrier and I can only get the mail out once the mail is put up. I had another carrier with me because we were doing two routes a day. The carrier and the customer had words. I didn't say anything because I feel like the customer is always right, and I just let the man talk. The carrier said something to him, and then she ran. I guess since he couldn't get to that carrier, he got to me. He swung me against the banker boxes which bruised me down in the lower part of my stomach, then he threw me to the ground and started kicking me in the stomach. I was taken to the hospital and once they took X-rays, they saw the blood. They inserted tubes into my stomach.

I was in the hospital for about 7 or 8 days, but the day I went into the hospital a couple of the carriers came to see how I was doing. The very next day I got a phone call from one of the officials. I couldn't even talk because the tube was down my nose and into my stomach, and he asked me when I was coming back to work. I hung up the phone on him. I couldn't believe a person would be so nasty. Not only that, I could have probably gotten a case against the doctor, but he came into the post office to complain that one of the carriers was rude. One of the supervisors said, "What do you mean rude? Where was this? What was the address?" and when he gave the supervisor the address, instead of calling the police to come and get this man, because they could not find him since he beat me up, the supervisor chased the man out of the building with a broom stick and the man filed counter charges against the post office. Somehow that threw my case out. My charges against him were reduced to a misdemeanor.

When I got home from the hospital, the post office called me everyday, not to see how I was doing but to let me know, "You know you know you only have 45 days" and I mean 45 consecutive days. They'll call and say, "You only have 20 days left, when is the doctor going to release you?" And that went on the whole 45 days. They called the doctor everyday to pressure him to release me to work. The doctor told me that he's so tired of them calling him to release me to full duty until, finally, one day he did. This was after he released me to restricted duty sitting in the office because I had a real bad time with this incident. But one day he said he was tired of the post office and didn't want to deal with them, or post office patients anymore. He said he didn't want to do anything for them because he didn't want to fill out their forms; he didn't want any of their business, which meant I could no longer be treated by this doctor. It was a while before I found another doctor, and he wasn't a very good one at that. Workers' compensation had stopped taking care of the medical bills and all other payments came out of my pocket, so I went back to work.

After the beating I started having a lot of female problems. I started bleeding most of each month. The most days that I went without bleeding was about 8 days during that time, the rest of the month I bled. I tried to find doctors to help me. They gave me medication after medication, but it didn't stop the bleeding. Finally, my girlfriend gave me the telephone number of her gynecologist. He gave me an ultrasound on my stomach and uterus to make sure that he didn't see any fibroid or blood. He saw a small fibroid in one of the ovaries and put me on birth control to try to regulate my periods. I stayed on the birth control for about a year, but I still bled. Finally, he said, "This is not normal," and scheduled me for out-patient surgery. What he found was scar tissue on my stomach that had attached itself to my female organs. He had to cut the hairline to go in to remove this scar tissue and to this day, I still have that pain in my stomach, and there's nothing that can be done. The doctor told me that this pain is forever.

POST OFFICE INJURY NUMBER THREE

"When the doctor finished treating me, I had on a cast from my hand all the way up to my elbow. You could tell the supervisor was angry. He said the doctor, "Well, can't she do anything...?
Valerie

When I went back to work I was dealing with both my shoulder injury, which never properly healed, as well as with my stomach problems as a result of the beating. The post office I worked in was very small, and very crowded because they had four zones to handle. We couldn't walk because of the tubs of mail; we had to climb over the tubs. At the post office every carrier that's a regular carrier has cases where we case flats.[1] We also have this shelf that looks like a bookshelf, but it's only a foot wide and about six and a half to seven feet tall. This is where the flats are stacked. The mail that cannot fit in the case are set on the very top of it, and the ones that could not fit on top were sat in tubs all around us; so we had no where to move. We had to stand in that one spot. I don't know who stuck a flat of banded JC Penny catalogs, the thick ones,

---

[1] A case is a box where the mail is put; a flat is a flat letter

you know, the Summer, Winter and Fall editions, up in the case. Anyway, it slid down, and I didn't have anywhere to jump so I tried to catch them. They weigh, I guess, at least 25 to 30 pounds, and they hit my right hand and bent my hand all the way back. I went and told the supervisor what had just happened, but I was ordered out on the street anyway. I was always told that when they give you a direct order you have to do whatever they say, and fight it later. So I went out and did my route, and when I finished I came back in, punched out and went home.

This was on a Saturday. I didn't even bother to go to the doctor. Of course, I didn't know at the time that I had fractured my knuckles, but when I woke up the next morning my hand was stuck in a cough position, and it was swollen real bad. I got up, got dress, and debated whether or not I should call in sick and go to my doctor. I said to myself, "No. I'm going to work and I'm going to tell them about this hand, and they are going to have to take me to the doctor today. I'm not going to take this mess from these people no more." So I went in to work and I showed them my hand. There were three supervisors there, and I said to one of them, "You ordered me on the street and I told you that those catalogs fell on my hand and bent my hand all the way back." I said, "Now look at my hand. I want to go to the doctor." As I was talking with one of the other supervisors, another one said, "You're not going anywhere, get back to work now!" When I didn't say anything, he said, "Do you hear me talking to you?" "I'm talking to someone right now." He told the supervisor, "Give her a letter of warning; give it to her now." And I said, "Give it to me. Give it to me. I'm going to fight it and I'm going to win because I told him on Saturday I got hurt and he didn't let me go. I went home, and today my hand is swollen and I'm going to the doctor." So he went into his office and told one of the supervisors to take me to the hospital.

The supervisor took me to the emergency room where they took X-rays. They told me they think I had fractures across the knuckles but weren't sure because my hand was so swollen. The doctor asked, "Why did you wait so long before coming in?" And I told them, "Because the supervisor at the post office, told me to go to work, and they did not let me come to the doctors." The doctor told him they should have brought me in when the accident happened because it's hard to see how bad a fracture is when the body part is swollen. The doctor put me in a splint and told me that I needed to see an orthopedic hand specialist. He referred me to the same doctor who treated my shoulder.

Anyway, the supervisor took me back to the office and called the orthopedic doctor and asked if he could see me that day because he wants to put me back to work as soon as possible. Well, they wanted me back to work so badly that when he got off the phone with the doctors office, he said to me, "Come on. The doctor will see you now."

When I think about that incident now, I laugh. At that time the supervisor took me to see the orthopedic surgeon, I just had a splint on my hand. But when the doctor finished treating me, I had on a cast from my hand all the way up to my elbow. I *really* couldn't work then. You could tell the supervisor was angry. He said to the doctor, "Well, can't she do anything?" He asked me, are you right or left handed?" "Right," I said. "Well you can answer phones." The doctor said, "If you want to put her to work answering the phone, her arm has to be elevated." I said to the supervisor, "But don't you think this is stupid? I can answer the phone, but I can't take the messages." And he says, "Well, you tell them to hold on, and you get somebody." It was absurd. I had the cast on for 6 weeks.

When the doctor's took the full cast off, they put on a smaller cast. With this cast I didn't have to have my arm elevated so they said, "You can throw mail in the P.O. boxes with your left hand." So they made me throw mail until my cast came off.

THE ENDLESS ROUND OF DOCTORS

"The pain started getting so bad I could barely see sometimes because it was ... so blinding, it made me dizzy...it started making me throw up ... and every doctor said they couldn't find anything ..." Valerie

I continued to go to the doctor because of the shoulder injury. I was still having the tingling which went down to the hand. I was referred to another doctor who turned out to be the biggest quack ever made. He said that I may have had a shoulder strain or sprain or something like that, and he couldn't tell whether or not I had a shoulder separation. The doctor did not take any X-rays of any sort, and the man barely touched me to examine me. When I told him about the hand he hit me in my wrist area. It hurt real bad and sent pain down into my fingers. He said he thought I had Carpal Tunnel Syndrome, and sent me to another doctor who stuck electric needles in me with electrical currents.

They also tried to do an EMG in the spine region, but it was so painful that they could not do it. They could not get a reading because they were afraid that I would move and break the needle off into my neck and spine. At that time, I mean the pain started getting so bad I could barely see sometimes because it was right up there; all on this right side, and it was so blinding, it made me dizzy. It started making me throw up because the pain got real bad and every doctor said they couldn't find anything. Worker's compensation finally authorized me to see another doctor.

The first time I saw this one particular doctor I was in so much pain. I had this swelling over my shoulder region, and I had the tingling, and I told him about it. I told him about the pain down my spine and into my legs and he said he could only do one thing at a time. He said, "Which one bothers you most?" "The shoulder and arm region." "Well, let's deal with that," he said. But when he wrote his first evaluation he said he thought that I had embellished what I said I felt, and this made me angry because the pain I felt was not only real, it was constant. Anyway, he decided that if I was not embellishing what I was telling him about the extent of my pain, the only thing that would cause that type of pain would be RSD or reflex sympathetic dystrophy. He sent me for nerve block which the anesthesiologist would not do because he did not feel it was RSD. He gave me local anesthesia to try and stop the pain, but the pain never stopped. The doctor sent me for another EMG because the tingling in my hand was real bad. I couldn't hold cups or anything anymore because I didn't have the muscles or the strength in my hand. The tests showed that I had Carpal Tunnel Syndrome which was getting progressively worse, so he sent me to a hand surgeon.

I told this new doctor about everything that was going on in my body, in the hand, in the shoulder, and in the spine. He asked, "How long?" "Since 1986," I told him. And he says, "Nobody's found out what's wrong?" "No." Well, the doctor wants you to have carpal tunnel release, and I won't do it until I find out what's going on with the shoulder because if I cut that carpal tunnel nerve, and it has something to do with the shoulder, you're going to be worse off." He sent me to see a sports orthopedic specialist. When he examined me for the first time, he was trying to rotate the shoulder. It didn't move, and he said that the shoulder had frozen and I may have a rotary cuff tear. When he examined me, the pain was so bad I blacked out. They used smelling salts to revive me. He had pushed the arm up, and because it hadn't gone up in so long, he forced it up, and it hurt so bad. All I remember is the smelling salts. When I woke up he said, "Gee, this thing is really bad. I can't do anything else for you until I get an MRI." So he ordered an MRI. This was the first MRI or X-ray on this shoulder since the shoulder injury occurred. This is now two years later.

At that time, I was already taking Elavil, an antidepressant, Darvoset, a pain killer, and Robaxin, an anti-inflammatory. Tests results showed that I did not have a rotary cuff tear but rather tendonitis, and a synovial cyst near the AC joint of the right shoulder which occurred from the fluids in the shoulder for many years. I went back to the other doctor and he said: "Well, since the MRI does not show a rotary cuff tear, I can go ahead and do the release, which was done in the Fall. I remember one day after I had the surgery I was in my kitchen when it started raining. I had my shades up. I opened the refrigerator, and I had my right hand on the door and as I turned to do something on the stove, lightening came through the window and hit my right hand. When the lightening struck, it jolted my arm. I had a little bit of movement in my shoulder after it happened; it wasn't perfect, but it was better than it had been.

I called the doctor and told him what happened, and he had me come in to see him. He was amazed, but he was still concerned and he wanted to start me on injections. The injections not only made me drowsy, but it brought out the pain worse than before. He said he had wanted to do a series of those shots but since it only increased the pain, he would not give me the shots, so he sent me to therapy. This was a year later. In therapy they had me lift weights that come over my head so that it could pull my arm up. But the pain got so bad that I couldn't do it. The next day I was swollen and couldn't move. I called the doctor, but he says, "You still have to go; there is going to be pain." So I continued to go.

THE EEO, GRIEVANCE AND POLICE

"I started crying because I couldn't get up and knock him out ... I just wanted somebody to come in here and beat his behind for me. They called the police on me ... She asked me, "Do you want to kill somebody?" I told her I just wanted them to leave me alone..." Valerie

During this time the post office was up to their old tricks again. They had given me an auxiliary route which consisted of two and a half hours of walking and two and a half hours of postal boxes. I did that for about 5 months. Then they took that route away from me and told me that they didn't have work for me anymore. They would let me come to work, and then they'll meet me at the time clock and tell me, "Go home, there's no work." I'm a regular who was hurt on the job. But these people were able to do anything they wanted. I had filed for worker's compensation for my shoulder, the beating that I took from the postal customer, and for my hand. Nothing was approved.

I bided for a route that was all driving and postal boxes because I got tired of them sending me out walking and putting that satchel on my arm. My doctor didn't want me to do any of that because I was only making myself worse. But I would case my route and put a lot on the street for somebody else. Then I would walk casings and do markup and stuff so I could keep myself busy in there so they wouldn't say too much to me. Everybody bided for a rider position. There was this one co-worker who I befriended. We would always talk about our marriages and things like that, always sharing advice and what not. We had a great working relationship; she knew what she was doing and so did I, and we worked well together. Well, she bided for a 204B[2] and got it, and I was assigned to her. Then she changed; she started acting different toward me. It's like the job went to her head or something. She began to harass me. I would come to work, and I know the guys in there—we all used to play little games on each other. So one particular morning, I don't know if it was a game or not, but somebody had taken my time card and I told her, "I can't find my time card." She said: "I want you off the workroom floor until you do." I told her, "That don't make sense. The time clock is on the workroom floor, so how am I going to do that." She replied: "Don't get me started." So I went and waited for a few minutes because I still had a few minutes before I punched in, and when I came back I found my time card sitting there. So I swiped in.

    I had to case the route all by myself that day. As I was casing the route, one of the carriers came over, and he started casing the letters. He said, "You go ahead and line up the flats." So I started lining the flats, and I had put some of them on top of a APC which we used as a table to line flats. The clerks were spreading mail, and I was trying to move some tubs out of the way. The 204B I was talking about came over and pushed the APC over to the other side of the room, and then said to me, "You need to go over there and get that APC." I said, "You need to pull it back over here because you could have given me just a few seconds to pull that APC in. You are delaying the mail." "No, you're delaying the

---

[2] Acting Supervisor

mail if you don't get it," she said. I left the flats right there and filed a grievance and an EEO.

The EEO took place before the grievance, and they asked how I knew her (the 204B). I told them that she was my router, and we had worked well together. I told them we talked, and I gave her advice about her marriage and stuff like that but, after she became a 204B it was like she was assigned to me to harass me. I told them that I was tired of that and that I couldn't take it any more. I told them that she pushed that APC out of anger. I could see if she said, "Let me move this, and you can get it when they go by, but she pushed it out of anger, and I said that I wasn't going to touch it." When she was called in to testify, she agreed that she would not do that anymore no matter who tried to tell her to do it. When the grievance came up, about three days later, they saw that I had won the EEO, and when I told them my story one of the manager's was upset, and said that if she (the 204B) was going to be a supervisor, she needs to go to training because there is a certain way to treat people, and that is not the way. So I won my grievance as well. But, about a week later I experienced a reprisal as a result of filing an EEO.

The 204B quickly forget everything that was said at the hearing because she started picking on me again. She yelled at me on the workroom floor, and told me that I was stupid. I turned around and looked at her and at that moment I knew I was going to read her good. I said, "Nah, I want to see you in the office." I told the shop steward to come with me because I wanted a witness. After I told her what I thought about her yelling at me and calling me stupid. I told her that I was going to re-open the EEO because they told her that she was not to treat me that way. She said she did not care what I did.

When we left the office I called another official who came back to the office and as I was sitting there with the shop steward doing my EEO papers another supervisor came over to us and said, "Punch her off the clock, punch her off now!" And I mean he didn't even say "Boo! He just came over and saw me doing the papers and told the supervisor to punch me out. The two of them went off to the side and spoke with each other, and when the shop steward returned he said, "Valerie, drop this EEO because I don't want to have to punch you off this clock. If you don't do it, I'm going to have to punch you off this clock." I looked in his eyes and said, "Then you are going to have to do what you have to do because I refuse to be treated like this. I'm a grown woman. I come in here, I got hurt on this job, several times, and you all treat me like an animal." He said, "I've never treated you like that." "But you are doing it right now. That man dictates to you what he wants done and you all do it. You don't have a mind of your own. You know I work hard, even with this pain that I suffer with every day. I come in here and do over and above what I should do as a person on limited duty." I told him, "I do your work and you sit there and you get the recognition for it not me." He said, "I don't want to do this." The other official came over and started yelling, "Punch her off," he kept saying, "Punch her off, she don't need to be on the clock, punch her off now." At that time I started crying because it hurt that I couldn't get up and knock him out. I wanted to go up and kick his behind, but I knew that if I did I would have to kiss that job good-bye forever. It ain't the job I wanted; it's because I got hurt on that job and these people owe me.

I cried. I tried so hard to stop crying, but I just couldn't. I was so pissed at that time that I just wanted somebody to come and just beat his behind for me. So I called my dad, and I said, "Daddy," "What's wrong with you girl, who's bothering you, what's wrong?" And I thought about it and said to him, "That's OK, Dad. It's OK" and I hung up because I knew that would be wrong, too, and two wrongs don't make a right. I tried to fill out a slip to go home, and my hands were just shaking, so I went and I sat in the bathroom. As I was walking into the bathroom, I walked along the wall, leaning against the wall, just rubbing my shoulder against the wall just trying to calm myself down because I was that upset. I went into the bathroom and was so sick on my stomach, I just started throwing up. I had not eaten anything so it was just foam. I got up and sat on the stool that was in there. I leaned my head against the wall, and I just cried. A couple of clerks came in the bathroom and then walked back out. They must have thought that I was crazy. The next thing I knew someone knocked on the door and asked if anybody was in there, and I said, "Yeah." She asked if anybody else was in there, and I said no. The door opened, and I was surprised to see that it was a police officer. They called the police on me.

She (the policewoman) said, "What's wrong?" "They just pick on me so much, all the time; and I just got so angry, I started crying. That was the only way to keep me from doing something wrong." I told her, "I cried because this job, they owe me because my body has been beaten down and I'm not about to mess up by hitting somebody and getting in trouble." She asked, "Do you want to kill somebody?" "No" I said, "Because if that's what I wanted to do, I would have done it. No, that's not in my heart; I just want them to leave me alone." So she said, "They requested that I take your belongings." "What do you mean my belongings?" "Your locker and your purse and your personal things." I asked, "You're not going to pat me down, are you." "No, I just want you to empty your pockets and everything." So I emptied my pockets. She said, "You need to hold up your pants legs." And I held up my pants legs. She said, "You need to pull your blouse up around your belt so I can see if you have anything stuck in your pants." And I did that. She said, "Now pull your shirt tight around you, so I can check for weapons." And I did it. She also checked my locker and my purse, and found no weapons.

She said, "Come on, I'll walk out with you." She walked out with me and told them, "I can't arrest her; she didn't do anything wrong; all she's doing is crying and if you people would leave her alone she wouldn't be crying. I can't even write a report on this." So she called in on her walkie-talkie and said there was nothing happening, and that she's on her way.

When they didn't get any action from that they called an ambulance, and when the rescue workers got there the supervisor told them they think I am dangerous to myself and everyone around me. The ambulance people questioned me and said, "Are you OK?" "No, I'm not OK. These people keep bothering me, and I'm so sick of it." "Well, why are you crying?" "Because it hurts that they are always picking on me." "Do you want to kill them?" I said, "No. Why does everybody keep asking me if I want to kill somebody?" He said, "Well, do you want to kill yourself?" I looked at him and said, "Are you crazy? I love myself better than I love myself." "Well, they want me to take you to the emergency room to get psychiatric care," he said. "But there's nothing wrong with me." "Ma'am, they are mandating that you go." So the supervisor came over and said, "It's a direct order." "Well, I want my girlfriend to take me." "We can't let you do that. You must go in the ambulance." So they took me to the emergency room in the ambulance, and when I got there, my girlfriend met me.

    The doctor talked to me, and he said, "Why are these people like this; there's nothing wrong with you? You're feeling better now that you're not in that place. You're not crying anymore." He said, "Take a couple of days off, and here's a note. When I called in sick the supervisor said, "You have to bring a note from the doctor." "I have one." "Is he a psychiatrist?" I said "No." "Well so-and-so- wants a note from a psychiatrist." I had never been to a psychiatrist and didn't know where to get one from. He said, "I guess look in the medical book from Blue Cross/Blue Shield, and see if there's one in there." That's what I did and located one nearby. I made an appointment for which he charged me $350.00. He said, "I don't understand it. Why do they want you to see me?" "Because they think I'm dangerous to myself and everybody else, and that is the only way I can go back to work, if you give them a note stating the diagnosis and prognosis and that I'm no harm to myself or anybody else." He said, "These people are so stupid." He gave me a note with the diagnosis and prognosis that stated that I was just upset at the time and stressed, and that I am not a harm to myself or anybody around me.

    The next day I went back to work with my letter from the doctor. As soon as I walked in the building and punched in, the supervisor came up to me and said, "You have to get off the workroom floor." "What? Here's my doctor's note." And as always, I made copies of the note. He said, "So-and-so told me not to let you back on the workroom floor." "I said, "I don't get it. When I called in and he told me to see a psychiatrist, I saw a psychiatrist. My note tells you there's nothing wrong with me; I'm not going to hurt myself or anybody, so now I can't come back?" "If I have to tell you again to get off the workroom floor, I'm going to have to call the police." "Well, I'm not going to give you a hard time, but can I see the union shop steward first." "No, off the workroom floor." Well, I have a big mouth and I yelled out the shop stewards name real loud and everybody said, "That's Valerie, she's yelling for you. The supervisor told her to get off the workroom floor." The shop steward came out and said, "What's wrong." "He told me to get off the work room floor; he won't tell me why; he said so-and-so told him to tell me to get off the workroom floor." I gave him the letter from the doctor. He read it and gave it back to me, and then the supervisor read the letter and said, "This gives diagnosis, prognosis, and that she's no harm to herself or anybody. She's supposed to be back to work." Well, so-and-so told me don't let her on the workroom floor. He's the boss." "We'll go and see him," the shop steward said. The supervisor then said, "OK, but we'll walk around that way so nobody can see us." So we had to walk around all the cases so we couldn't be out on the floor, and we went into the office. But so-and-so would not accept the doctor's note. He said, "No, I want to send her to our own psychiatrist, and that's the only way she's going back on the workroom floor." The shop steward said, "So you are going to pay her administrative leave?" "No. If she wants to use her annual or sick leave, she can."

    A month later I received a letter to see this doctor, who I later learned was not a psychiatrist but rather a psychologist. He made me take a written test; it was like 800 questions, and they were the same questions over and over again but rephrased, asking me would I kill somebody; if somebody attempts to hurt me would

I try and kill them; if I was in danger would I protect myself; you know, questions like that. I answered them to how I felt, and I'm not a violent person, but I ain't going to let you hurt me either. He spoke to me for about 30 minutes. He refused me when I asked for a copy of his report. I had to order it from the post office. I called the union and told them I saw the doctor, and I wanted the union to order the paper work. They said OK, and I had to sign a release.

By the time the post office called me back to work I couldn't walk because I was in so much pain. I couldn't get off the bed because the pain had gotten so bad in the spine area that I had to be lifted and carried around. My father would come over to my home and lift me up and carry me back and forth to the bathroom or whatever I needed. My mom would help bathe me and they would take my kids back and forth to school during this time. My husband and I had divorced by this time so I was alone and needed my parent's help. I was so frustrated because before I went to work for the post office, I was healthy and could always do for myself. Now after three or four job-related injuries, I was now reduced to throwing myself off the bed and crawling to the bathroom.

I would get dressed while sitting on the floor. I recall this particular day I had to go out and pick my kids up, and I had a fever of 102. I tried taking Tylenol to break the fever, but it did not work. I had to take my kids to their doctor this particular day, and their doctor told me that I needed to go to my doctor, which I did. He sent me to the hospital to be admitted right away. He said, "I don't know how in the world you are trying to walk or even drive when you can't even sit up." While I was in the hospital they did a spinal tap and took other tests. My shoulder was in excruciating pain. They hooked me up to an IV with pain killers and antibiotics. I stayed in the hospital, that time, for seven days.

It took me a month after I got out of the hospital to recuperate. It took me that long to get up and walk around. I called my job and told them I would try to come to work and they said, "You can't come back to work until you get a note from your doctor saying that you are not contagious." I told them that I did not have anything contagious, and they said, "You still have to have the note." No problem, I got the note and took it in. The doctor said that I needed to work inside because I had to try and gain my strength in my spine and back area. I worked in the office for awhile and then I started getting stronger. I was sent back to the streets to deliver express mail. When I came back in, I had to do labels on the carrier cases and vehicle cards and other sundry type things.

Management sent me a little note typed on a Form 13 informing me that beginning, I don't remember the date, my start time would be 1:00 a.m. ending at 9:30 a.m. At that time I couldn't reach above my shoulders, and I am right handed. I asked them, "What am I going to do at that time of the morning? They said, "You have to case." "I can't case; I'm right handed." "Well, you'll have to learn how to case with your left hand." I did it because I needed to work. I was single at that time and I had to do what was necessary. But it took it's toll on me. I was taking a lot of medication at that time, and I was also going to physical therapy. Now that my schedule had changed I couldn't take all the medication required because the Elavil made me real tired, so I'd have to take that at bedtime. When I got off at 9:30 in the morning I had to take just one of the pain pills and lay down to rest until it was time to pick the kids up. When we got home, I fed them, they had their baths, did their homework, and watched television. At 4:30 I would lay down to rest, and at 8:00 I would get back up to get the children to bed. Then I had to lay back down and sleep until about 12:00 at which time I got up, showered, and went to work. I did not have anyone to stay with my children at that hour, my mother couldn't do it anymore, and I hated leaving them by themselves like that.

After all my years of service with the post office, this is what I have been reduced to. I went into service a vibrant healthy, and happy woman. I retired on disability with a broken body. Oh, I've won all my grievances, EEO cases arbitration cases, my retirement disability, and even my social security disability cases.

But it does not begin to make up for the broken body I now have to live with. And I'm still fighting for my worker's compensation for my job-related injuries. They owe me that because I still live with pain in my body everyday, all because of my job-related injuries.

I am giving this interview because I want the readers to know about the mistreatment of injured workers. When you are hired by the postal service, management paints a pretty picture about benefits and working conditions, but when an employee is injured, management wants to put them out to pasture to be shot like cattle.

CHAPTER VII

"It feels to me like the government is more insidious ... I'm not quite clear what the post office is doing, what their motive is for treating their employees the way they do ..." **Dr. Tai**

I was in college in Greensberg, North Carolina during the sixties, and this subject of management abuse reminds me of what the FBI did during those times in terms of harassment. The government feels that they have a license to do whatever they want to do. An example of government power is the recent situation at the Olympics where they accused this man of the bombing. They were wrong about his involvement in the bombing, yet the government refused to apologize for their treatment of him. So, it's very interesting, you know, about the power of the government.

As a social worker with a masters, I have treated a number of postal workers, but only one for an extended period of time, and based on my treatment I have reflected on this theory of management abuse. Many of the postal employees are war veterans, either from Vietnam or from World War II. If you look at the psychology of war, in terms of feelings and emotions, there's a lot of mistreatment. There's a real horror about war but soldiers are not trained to pay attention to the war itself. In basic training you are taught to become a part of the team so that you think like a group, not as an individual. It's not about paying attention to personalities and feelings, and that kind of stuff. What soldiers go through is a lot like abuse. But it's not called abuse, it's called basic training. So when these soldiers come out of service, particularly those who returned from the Vietnam War, many of them really couldn't deal with the world they returned to. A lot of them are just trying to survive.

So now, there are some veterans who go into an organization like the U.S. postal service, become supervisors and managers, and there's abuse there. Only it's not seen as abuse, and because it feels okay, it's not dealt with as abuse. I think for anyone to recognize abuse they have to have to have some sensitivity to their own abuse.

I was really struck by a television journalist who did a segment about taking some World War II veterans back to where they had fought the war, I think it was the Cliffs of Dover. It was very interesting to watch the soldiers; they were fine until they hit land and walked the grounds on which they fought. As they remembered their personal experiences, they related to the realization that they had not dealt with their wartime experiences at all; the horrors of the war. Now, the reason why I'm talking about this is because in my profession I deal with those buried emotions. Just because emotions are buried doesn't mean past experiences do not have an affect on us. We don't know our behavior if we're not sensitive to it, but rather we act out these feelings in an unconscious way. The feelings go from conscious to unconscious, so we act them out. So if we're in denial we don't necessarily know what we're doing. And if we're not acknowledging abuse within ourselves, then we're not necessarily going to acknowledge abuse from other people. What I've heard from my clients about the behavior that goes on in the post office in terms of management's behavior, now I characterize that behavior as abuse. The abusive behavior is not acknowledged and it seems like managment says, "Well, this is just the way we operate."

There is a piece that I inevitably do with some of my clients because many times they've been abused and they've been harassed, and *they* don't even acknowledge it. They suffer with symptoms—they're depressed, they're anxious, they can't sleep, they can't eat, their back goes out, and a lot of that has to do with stress and emotions. In working with them, one of the things that I help them see is that abuse is not okay. First, they have to identify it. I help them see that it is not okay to be abused. What I wind up having to do over time is begin to gently lay it out to them, and say, "Well, there are other alternatives." I help my clients to see that a particular tone of voice, the manner in which they're spoken to, is not an okay thing. They are having feelings about it, but they think that they are crazy; they say, "Well, no one else sees this so I must be crazy." Well, you are not certifiable because you don't recognize it. You must look at all of the symptoms, all the stress, and all the things that go on in your job and your personal life. You look at that total package. That's how I help them identify what's going on.

What I've seen is a lot of depression, but many people when they come in to see me don't even know they are depressed. They don't know why they are acting the way they are acting. All they can say is, The job is very stressful." But what is the stress? They tell me, "I'm not sleeping. I'm not eating. I'm very nervous. Things have changed. I don't understand why things have changed at work. I'm being treated differently. I feel like I'm being set up, and I don't feel like I'm paranoid but maybe I'm acting like I'm paranoid because it feels, you know, it doesn't feel right. People that were friendly with me at one point are not friendly with me now. I thought I could trust them, and I can't trust them, and I'm missing some signals." Those are the kinds of things that I hear from my postal service clients.

Right now, I'm thinking about one person in particular whom I treat. It's incredible. A manager will see an employee that may not be in management and see their potential, and guide them along and encourage them, and really give them a lot of positive feedback because they think that they can guide them along the party line. So now the employee is encouraged, and puts more than 100% into their job because of the strokes, the encouragement.

They are putting out more and more, taking on different tasks in addition to their own job, and in many cases this causes stress for them, but they continue, nonetheless, to put out more and more. And then, all of a sudden, when the employee starts asking questions, all of a sudden management is not recognizing or encouraging the employee anymore. Management starts pulling back and does a 180 degree turn on the employee. Now, management starts finding fault with the same employee who, just yesterday, they stroked and encouraged. Now they start criticizing. Maybe that employee was in line for a position, but management now gives the position to someone else. And this cracks the eggshell.

It's the eggshell theory. An employer takes his employee as he finds him. What I mean by that is, an employee may have experienced some childhood, or adulthood, trauma, but has dealt with it or, is dealing with whatever horrific experience it may have been. They are working, and doing fine. But when they are placed in this kind of situation that I just described, it's like management dangling a carrot right in front of a horse. And then, all of a sudden, the carrot is gone. Now, something happens to the employee. After being told, "You were wonderful, but now you're lousy." That really is messing with someone's mind. What happened? Why is it that one minute management is saying to the employee that they are doing excellent, and giving the encouragement to do the job for the promotion, and then the next minute start criticizing. Management may even say to the employee, "Don't worry, you've got this position," and then turn around and give it to someone else. It's political, and that's not acknowledged.

In terms of the emotional impact on the employee, there's devastation, confusion, and there's questioning. They are now beginning to question their *own* ability, and depending upon their personal makeup, some people become very hostile and defiant. This behavior, however, just plays into management's hand and they say to the employee, "See, you're not the kind of employee we thought you were."

This goes for both people who are emotionally strong as well as those who have, in the past, not been so emotionally strong. For someone who's not emotionally strong, the devastation takes more of a toll. But it also happens to people who are even stronger. People who are even stronger have come in, they can't sleep, they don't know why they're not sleeping, they're not eating, or they're eating, they're overeating. They can't relate to people, things are changing, and some of them are having nightmares. It's just real stressful, because now they're placed in a situation where they are trying to perform their job and the emotions are impacting them, and they can not perform on the job, so they're confused.

It's not so easy to say, "Get over it." There are long term effects. If a person has been in therapy for other things, they are going to be in therapy a little bit longer. I can't begin to tell you the number of people that are coming into therapy now because of the way they are being treated by their employers; they are really being treated very, very badly. Basically what management seems to be saying to them is, "We don't want to hear what you're going through, we just want the bottom line—performance."

My experience with treating postal workers has given me an opportunity to compare the treatment of these employees by their supervisors with employees and employers in private industry. I have found that treatment of employees are done in a very different way. It feels to me like the post office's treatment of its employees is more insidious. I think with big business, with merging and downsizing, they treat employees differently; perhaps from an economic standpoint. I'm not quite clear what the post office is doing, or what their motives are for doing what they do to their employees. I mean, I see people with disabilities working at the post office, and it gives you the impression that the government is good in terms of working with people with physical disabilities. And they may be, but in terms of dealing with employees and their emotions, I don't see that those problems are taken into consideration. I hear that the postal Employees Assistance Program (EAP) is trying to do something in terms of the employers. I'm referring to administration, because from what I see the administrators really need some training. They need on-going training in terms of how they sensitize themselves to the treatment of their employees.

I think of the employees that have wound up being so angry, and so disturbed that they have ended up shooting others. My question is, how were they treated? What precipitated that violent behavior? Certainly, some of them have their own personal stuff. Everyone has personal stuff to deal with, but going into a hostile office environment, to add to that personal stuff, is not very helpful. So one of the questions that I always have to ask is, what was the quality of the treatment on the job? I really question management's role. I don't know whether the government is saying. "Let's take a look at how we treat our employees. Can we treat them better? Can we can do some preventative work rather than doing clean-up?"

People that are happy go lucky all the time, I would be very concerned about that behavior too. I ask, "What signals are you missing?" I mean, if he's happy-go-lucky, is he a happy-go-lucky loner? Because one of the profiles I've seen of one of the people that has done some shooting at one of the post offices was that of being a loner. People do have all sorts of facades, and in that particular instance there may not have been anyone that he could really talk to about what was going on with him. Many times when your world begins to fall apart, if you're losing your family, and if you're losing your job, or things aren't going well on the job, they may not have learned how to deal with stress. Maybe the way they deal with stress is to act out; to eliminate that which is causing them stress. I think that maybe what post office management might want to look at is how much stress they put on their employees, and what causes them to act out. But if all they are concerned about is the bottom line, then all they will get are more complaints. I think in some ways it's good people are saying, "I don't want to be treated that way", and, "You can't treat me that way."

One of the things many people don't recognize in dealing with feelings is that our emotions get locked into our nervous system, and our muscular system. This places stress on our body; our body becomes very tense, and that can cause lack of sleep, it can cause the back to go out, it can cause distress in the stomach, distress in the bowel, extreme headaches, those kinds of things. Certainly, one of the things you look at first are the physiological manifestations; go to a medical doctor, and get checked out to see what's going on. The doctor might say, "Well, it's stress." So now you look for ways to deal with your stress. There are all sorts of relaxation, and therapeutic techniques now to address stress. But sometimes the person is trying to hold up stoically against the pressure. If there is no outlet for the person to talk about the feelings, and if the employer doesn't provide that outlet, or if the person doesn't know how to deal with it, or does not know where to go for help, what happens is that he feels cut off. Sometimes the tension is held in your jaw, or you may hold it your eye. Some people have eye stress, and jaw stress and will go to the doctor and say their jaw hurts from clenching their teeth, or grinding their teeth at night. The stress is there, but they have not found a tool to release it.

Many times the stressed employee find themselves doing what their employer says, "Deal with it" or "Bear with it." It's not only employers, but a lot of people in our society say that. So there's all sorts of ways that people will try to compensate physiologically when placed under extreme stress. Areas where the stress is held in, like the back, neck and shoulders, eventually give way because those structures just can't handle an overload of stress. For example, you hold tension in your lower back, and also your shoulders; it's like you'll carry the weight, you'll bear up, and you tense up. But it's not something you just consciously think of doing; you just learn how to compensate in many ways. So if you're under a lot of stress and you are not relaxed, and you are asked to pick up something like a heavy mail bag, it just adds to the stress. In order to pick that heavy mail bag up, you tighten up, and you say to yourself, "I've got to lift this, so let me bear down." But, given enough stress, your body's going to go.

THERAPEUTIC WAYS TO DEAL WITH STRESS

There are different styles of therapy. The training that I've had over the years deals with how your emotions affect your body and vice versa. There's a therapy called bio-energetics, which really looks at the muscular and nervous systems. Your muscular system deals with how you armor yourself, which is what I was talking about earlier. That starts from infancy. Generally, we are very relaxed, but when we experience repeated stress we hold our body in different ways. Today, there are different kinds of therapies and massages, like Roulfing. This particular therapy was developed by Ida Roulf. Her technique specifically deals with the connective tissue, which relates to our muscles and bones. With all the different types of massages, for instance, trager, shatzo, and Swedish massages, there are alternative ways of relieving tension.

A massage therapist, like myself, can tell you that when you start relieving the pressure in a particular area of the body, there is an emotion that is also released. That released emotion allows you to directly pin-point a particular incident, an emotional incident, an actual incident that happened and has not been resolved. That emotion is held in a particular part of the body. Many longer term therapies deal with these emotions in terms of what the injuries are, and what things are held inside that can be released. It takes a lot of energy to hold these emotions. When that happens, the person is not functioning at their top capacity. When the tension is released, and the issues are identified, people find new tools for dealing with those unresolved issues. Now they find that they have more energy to live a much healthier life.

What people are really learning, however, is how to relearn. When emotions get locked into the body, they get locked into the nervous and muscular system. In therapy, people can find out what those issues are, and learn techniques to deal with them through processes that, hopefully, will allow them to live their life stress free.

CHAPTER VIII

"When a person is deliberately messing with your mentality, insulting you, and embarrassing you and on top of that threaten you, the only thing that would come into any human beings mind [is] to kill ... I can see why people went to the post office and blew a supervisor away ... the post office will drive a person to that ..." Arnie

I started as an LSM clerk in a post office that had the hardest scheme in America. I worked in a post office on the West Coast and I think we had over a thousand addresses to remember. I chose the on-the-job technician position where you train people to study their scheme because I liked helping people. Basically that's what that job was all about, helping people. Unfortunately, I had been assigned to a station with a supervisor who had some animosity towards me. I'm a people person. I get along with just about anybody that gets along with me. I tried to get along with him, but he still knocked my fence down, and he just didn't try to get along with me at all. It was my understanding that he was transferred to that station because he had sexual harassment charges against him at another station, and to escape the charges, basically, what management did to cover that up was they moved him to another station so he could start his slate off clean. But unfortunately it didn't stop.

For some reason he had some type of attraction to me. It was as though he were possessed with the way I attracted both men women. It was my whole personality; he was attracted to me. He started with little digs, like, "Arnie I know that I can get more women than you," "Arnie, I know if I was your height I can get more women than you," "Arnie, how are these women in bed, I know you went to bed with them, what do they look like naked," "Arnie, give me some information, I want to know."

There was one particular female supervisor that I didn't know he was attracted to, but the supervisor was attracted to me, not sexually, but as a friend. I had a new baby and a wife with whom I was very happy. Anyway, he noticed that she and I were friends. Well, he had been asking that supervisor out for a date for a long time, and she had always refused him. So when she would come by my work area it pissed him off. It caused tension between me and him because I didn't indulge in his sexual conversations, and he just built up this anger inside of himself toward me. I tried to resolve this problem and tried to make him feel secure about himself because he's a short guy, and I'm not trying to be facetious or anything like that, but he was *much* shorter than I was. I was trying to get along with him, trying to give him some confidence about himself, you know, build his self-esteem. But I saw that those little conversations were not helping at all; it only stirred up more anger in him. Part of the problem was, it wasn't just *that* particular female supervisor that he was attracted to, it was any women who were attracted to me. As I saw him getting angrier, and angrier I knew that there was going to be a problem. I knew that my job was going to be lost somewhere down the line.

This supervisor continually harassed me, sexually harassed me, he disrespected me, humiliated me in front of my co-workers and peers, he even threatened me in front of my co-workers and peers. This man started spreading rumors about me. New supervisors who had just met me already knew about me. It was as though this supervisor went to my file and looked up every little detail about me. It upset me, but I couldn't do anything about it because this man had the power to do that. This man had a goal, and that goal was to get rid of Arnie.

Eventually, he trumped up charges against me and accused me of threatening him. This trumped up charge ultimately led to my removal from the postal service.

On June 23rd, this supervisor came to me with a charge that I threatened him on May 30th. Now, that is a very serious charge, and everyone in the post office knows that such a serious charge is cause for instant removal. What's funny about the whole thing is, if I was supposed to have threatened him on May 30^{th,} why did he wait so long to bring charges against me? When I was called into the office, an Inspector was there. I asked him, "If this is a serious charge, why wasn't immediate action taken against me." For some reason, he couldn't answer that question. This supervisor had been harassing me by disrespecting me, humiliating me, and by sending me to the EAP which he knows is for drug and alcohol abuse. He would put me on restriction, he tried to say I was late, and he tried to force me to sign documents that I was late. Anything that he could think of to have me removed from the post office, he did it. Finally on June 23^{rd,} he gave me a Notice of Removal. I guess he figured that I was going to be hostile and throw everything around and attack him, but no, I just accepted the Removal and went to my place of assignment. I still reported to work as normal but I guess that wasn't enough for him, because when I didn't act the way they wanted me to act, he used other little tactics like separating me from the work group. He started winking his eye and telling my co-workers, "Yeah, I got Arnie. July 1st is his last day,"

Finally the stuff hit the fan. It came after lunch one day. I guess maybe he had a couple of drinks or whatever, I don't know what the case was but that's when he started yelling and screaming at me for no reason at all. He tried to embarrass me by taking everyone's disk but mine. He then stood on top of a machine and he yelled over to me, "Arnie, bring me your disk!" Simply, I said, "Mister why didn't you take my disk along with the group?" and he said, "Arnie never mind that, bring me your disk." One of my co-workers came and got my disk and brought it to the supervisor and said, "Leave the guy alone." But the supervisor insisted, "Take this disk back to Arnie and I want Arnie to bring me the disk." Finally my co-workers calmed him down and he went on to something else.

As I was keying in the mail, this same supervisor approached me and said in a very disrespectful tone, "Arnie get up." Then he raised his voice, "Didn't you hear me say get up? Get up Arnie." At that point another supervisor came over and told me that I wasn't assigned to the seat. I simply got up, and said to my supervisor, "Mister, speak to me with some respect." He said, "Yeah, I'm going to give you some respect." He walked over to the phone. Next thing I know the postal police came up to the work room floor. He said the supervisor had called and said that I had threatened him. But little did he know the postal police had documented what the supervisor stated to him. Then the supervisor changed his story and told the postal police, "Arnie didn't threaten me he just was not doing his job."

When it came time to file a grievance the supervisor told his story, I told my story, and when the postal police told his story about the supervisor, he said, "That man is lying. I was the first officer on the sight, and I took his statement. He did not say anything about a threat."

The postal police was subpoenaed to testify at my unemployment hearing. He, and several of my co-workers who also testified on my behalf, explained to the Judge just what the supervisor said when he threatened me. Based on all the evidence presented, the Judge rendered a decision in my favor. Unfortunately, when the post office received that decision they reprimanded the postal police for telling the truth. Rumors began to circulate around the facility that the postal police and I were best friends, that we were calling each other on the phone, that I was over to his house and he was over to my house. Among other things, it was learned that his phone was tapped. Management put him through so much that he eventually had a nervous breakdown. I told my EEO attorney what was going on and she advised me what to do, which I passed on to the postal police. After that, they stopped bothering him.

The were several witnesses at both my unemployment and EEO hearings who testified that this supervisor said he was going to kick my butt. I don't know what he was trying to prove, but he said it *in front* of my co-workers. It not only embarrassed me, it hurt because of the simple fact I couldn't do anything about it; and if I had, I would have lost my job. So I just swallowed my pride and did what I had to do. Several of my co-workers were worried about my safety, so they waited for me downstairs until I punched off the clock and they walked me to my car.

But before I punched out, the supervisor came out of the supervisor's suite, and I guess to reinforce his earlier threat, he said, "Arnie I hope you remember what I said, I know which way you walk and I'll be downstairs waiting for you." And sure enough, he was downstairs when I came out of the building, but two of my co-workers were already down there and I guess when he saw them he just walked in place and tried to play it off like he was my friend.

THE REMOVAL

On July 1st I was removed from the postal service. While I was working at the postal service I held a second job and was grossing at least $60,000 dollars between both jobs. When I was removed from the post office, I suffered a serious pay cut, and bills were still coming in. I was still trying to fight to get back into the post office, unfortunately I didn't get back in. That caused a lot of pain because not only did I just buy a new car, my wife and I had a new baby; I had a family to support, I had bills coming in. This thing really hurt me. I felt like my manhood, my leadership for my family had been taken away from me.

I was upset. I began ignoring my family, I was ignoring my responsibilities, basically I was just laying in the bed like a bump on a log. My mind was just wandering, I didn't know what to do, I didn't know what to say, I didn't know where to go. I was mentally stressed, but I didn't want to go to a psychiatrist because I did not believe in psychiatrists. The first thing that comes to their mind is, 'you're crazy.' Seeing a psychiatrist has always been a taboo with a lot of people, including me.

But finally, one day, I must have been at the lowest point in my life, I was watching this television program about people who had tried to kill themselves, but did not succeed. This program caught my attention because that's what I had been thinking about. A guest on the show said he had tried to commit suicide, and suggested that if other people were having the same thoughts, they should first find a doctor and talk it out. I had my pride, but that person made me think that I should get help first before I killed myself because if I kill myself ... now I'm starting to talk logic here, *[Arnie laughs]* if I kill myself, the post office automatically won.

So I looked for a doctor. I'm going to be honest with you, it's hard looking for doctors. I found a doctor that I felt was just taking my money, and that alone almost discouraged me, but I continued to see him for awhile. I also kept talking to my attorney because I felt that she was the only one that really understood my situation, I guess because she was handling my case; her and my shop steward, they really worked themselves to the bone. Finally, my workers comp representative recommended a psychologist. I explained this whole situation to the doctor. I told him about the sexual harassment, the disrespect, the humiliation, and the embarrassment that I went through.

Even when I was trying to get a job, these people at the post office kept me from getting a job to support my family and maintaining the lifestyle which me and my family had become accustomed to. When I got fired, of course, I was looking for a job, and today most jobs require references. I used the post office for reference, unfortunately, even though it is not legal, they were writing harsh and mean things about me. They were convicting me without a trial. They would say that I disrespected and threatened a supervisor and, of course, with that kind of information no one was going to hire me.

I felt pain, I felt mental anguish, and that supervisor put that pain there. He orchestrated my removal and succeeded in getting me fired from the post office. It got to a point where every night, and I mean *every* night, even until today, I wake up and my teeth are gritted down, you know, crunched down, and when I wake up in the morning my jaws are so sore because I'm going through things in my sleep. I had never been through that before in my life.

I filed an EEO complaint because of the removal. I had all my witnesses lined up including the postal police who testified on my behalf at my unemployment hearing. At the hearing he stated that the reason he was testifying in front of everyone was to let them know that the post office has treated him like a criminal. All he did was tell the truth and they tapped his phone, they reprimanded him, they harassed him, they kept calling him to the office for minor things, and they hindered him from doing his job. All of that because he told the truth at an unemployment hearing. It hurt me. I felt responsible for him being written up. I was so upset because all he did was tell the truth. There were other witnesses who suffered retaliation by management as well, but this guy is a figure of law enforcement and for him to be abused like that really hurt me.

But isn't it funny that the unemployment Judge found the postal police and all of my other witnesses to be credible, but still, I didn't succeed in the EEO forum because *they* found the witnesses of the postal service, not my witnesses, to be credible. I have appealed the EEO decision.

POST OFFICE VIOLENCE

When I was removed from the postal service, I was thinking about getting me a gun and going to the post office and blowing all those supervisors away. But you see, that's where my faith in God came in. And I'll be honest with you, I got down on my knees because I didn't want to fall down into that category. I didn't want to be a statistic like that. But believe me I thought about that everyday. It's not hard to do, because when a person is deliberately messing with your mentality, insulting you, and embarrassing you, and humiliating you, and on top of that, threatens you physically, and threatens your livelihood, the only thing that would come into any human being's mind, is to kill. I pray to God to this day, and I am blessed, because I didn't resort to that.

You know, many people think that the postal employees are drunks, alcoholics, and drug addicts. But I say, "Look at me, do you see a drunk, or do you see a drug addict here?" In a sense, I kind of protect the post office. But being that everything I said here tonight happened to me, I can see why people went to the post office and blew a supervisor away. That may not have been the right way, but the post office will drive a person to that. You can be the sweetest person in the world and that's what would come to your mind if it ever had happened to you.

CHAPTER IX

"How could anyone become so angry that they would allow themselves to lose total control where they would arm themselves, come to work as though nothing were wrong, look for their supervisor, or manager, point a gun and shoot. Surely this had to be a sick individual. That's what I used to think. Little did I know, I too would be entertaining such thoughts. I knew I needed to get help, or risk losing it. That's the state of mind the post office reduces you to." Gabriele

I began my postal career some 15 years ago. Most of those years were spent in management. I earned my way. There were times when I witnessed upper management being abusive and discriminating, and employees choosing not to say anything because they felt if they spoke up, management would retaliate. I could never understand why they felt that way. That is because I was never directly involved.

I would hear of violence in the workplace, and I would think to myself, "What would provoke an individual to take someone's life, and shoot innocent by-standers? How could a person harbor such hate? What drives a person to carry a gun into their workplace, and shoot their manager? Surely this had to be a very disturbed individual, and more than likely, they probably had a troubled childhood." I was soon to understand the reasons for such violence.

Out In the Open...The Abuse of Injured Postal Workers

In 1994, I held a position as Supervisor Customer Service in a post office facility on the East Coast. I would come to work, and perform my duties with pride. I loved what I was doing, and at times, I was praised by the higher management for doing my job so well. The office I worked in consisted of several employees, including my manager. He was always talking about women, and how he loved Nordic women, in particular. He always joked with everyone, in particular, the female staff. I was among two women who reported directly to him. The other woman was very pretty, and much younger than I, and my manager obviously displayed a liking for her. This was clear from the preferential treatment he gave her. She soon became his personal chauffeur, taking him to and from the airport, leaving together to attend meetings, etc.

I was the senior staff member in the department. I was there the longest. Even before the manager was selected, I was the highest level staff member — with the exception of the manger. And, I was the oldest member. As senior to the manager, I was expected to be the one to cover for him in his absence, especially since the other woman who reported to him did not work in our office, but was from another area. I began to ask for training that would permit me to perform my duties more efficiently while covering for my manager. At first, my requests were simple ones; learning particular reports that he was responsible for. Simple requests were satisfied, but when I began to ask for more cross training into other departments that reported to him, my requests, although not denied, were just conveniently never addressed. Instead, he selected the younger woman that he was attracted to, to be trained. He began detailing her to areas that I had requested. It was obvious to the entire staff what was going on. But, no one said anything. I, however, decided to meet with my manager's boss to address my concerns.

I met with his boss, and after voicing my concerns, he instructed me to go back and talk to my manager once more, and then follow up with a written request. I followed instructions, and again met with my manager. He ignored me the whole time I spoke with him. He eventually looked up, only once, to say, "Don't worry about it." I handed him my written request that same day, and faxed a copy to his manager. In the letter, I repeated my request, stated my disapproval, and accused him of obvious favoritism and discrimination.

The next day my manager was silent. It was pretty obvious that he was upset over the fact I not only questioned, but had the nerve to send a copy of my letter to his boss.

From then on, things began to go downhill for me. I was now considered a trouble maker, and I was definitely on his sh-- list. Forget training. He would hardly speak to me. I knew he was plotting something, but I had no idea what he had in store for me.

THE GOLDEN RULE — FOLLOW PROCEDURE?

Approximately one week after I wrote my letter, a Limited Duty[1] clerk from our associate office, suffered a job-related injury. This happened early in the morning. He reported the injury to a supervisor. According to what the injured employee told me, he hurt his back while moving about in the office. I received his call at approximately 4:00 p.m. He explained to me that he immediately reported the incident to a supervisor, but the supervisor did nothing to assist him. He went on to explain to me that he received no assistance the entire day, and when it was obvious that the superviosr in charge was not going to assist him, he decided to call me. He told me he felt that perhaps the pain would subside, and even go away over night; but he just wanted to report the accident to me. I instructed him to seek a supervisor, and have the accident written up. However, he told me that he felt there was nothing to worry about. He said if he felt any different, he would call me the next morning.

The following morning, the employee left a message on my recorder explaining that the pain did not go away, and at this point, he felt that he needed to seek medical assistance. I called him immediately. After a short conversation, I walked into my

---
[1] Someone who suffers an on-the-job injury

manager's office, and reported the entire situation to him. After discussion with him, it was agreed that I was to leave the office, travel to where the employee was, write up the accident report, and see that he gets medical treatment. Before doing this, however, I spoke to injury compensation, and reported to accident to them. I then left to assist the employee.

From the time I first spoke to the employee the previous day, until the time I received his call on my answering machine the following morning at 8:00 a.m., and the time it took me to travel to his location to get him to the doctor, only 18 and a half hours had gone by since I was first made aware of the accident.

According to section 821.31, the general requirements of the E.L.M.,[2] the manager, or supervisor of the employee, or operation involved, must report all accidents on Form 1769, Accident Report, *within 24 hours* of the date of the accident, the diagnosis of illness, and the date the manager/supervisor was notified of the situation.

Since I had other administrative duties to attend to, I called my manager and told him I would be spending the afternoon at the associate office, and would not return until the next day. He responded by saying it was okay.

---

[2] *Employee Labor Relations Manual*, Form 1769, Accident Report

THE REPRISAL

The next morning, upon my arrival to work my manager asked to see me. When I entered his office, he didn't even look up at me, but simply said, "This is a pre-disciplinary meeting. I am serving you with a letter of warning for failure to perform the duties of your position, specifically, not reporting an accident immediately." I was devastated.

Perhaps it is hard for someone to understand how I could be so affected by a Letter of Warning. I explained to my manager that I reported the accident within the 24 hours as the E.L.M. instructs. I explained to my manager that I could not force an employee to seek medical assistance, nor report an accident. However, my words fell on deaf ears. My record was sterling. I never experienced any disciplinary action. Little did I know that this was just the beginning of what would soon be a series of events that were carefully orchestrated to remove me from my position.

I left his office, went into mine, and broke down. I can never explain why I fell apart, however, I did. I called my close counterpart, a dear friend who I always shared my joys, and disappointments with. She could barely understand what I was saying; I was so overcome with emotion. I just kept crying. She eventually talked me into calming myself down, and together, we came to the conclusion that he would never have threatened me with the Letter of Warning, had I not sent the letter to his manager charging him with discrimination. We knew what my next move had to be. I had to file a complaint.

I called the E.E.O. office.[3] This was the first time that I had ever filed an E.E.O. complaint in my 15 years of service with the post office. I felt so uncomfortable with the idea, but I had to report this abusive manager to someone. I was informed that I had to get permission from my manager to go to E.E.O., on the clock. I thought to myself, "This is getting worse." I finally got up enough nerve to call and tell him that I needed to go to E.E.O. He asked, "For what?" I told him that I just needed to go. He told me, "Go ahead." I went to E.E.O., and explained the entire story to them. They seemed to think that the threat of a Letter of Warning was unjustified. Like myself, they seemed to think that I had 24 hours, not 18 hours, to report an accident.

I began to feel sick. By now, I was feeling like my career in the postal service was over. I felt that once I made this move to file an E.E.O. complaint against my manager, I would be labeled a trouble maker and I would be black-balled. I had always heard employees speak about the dangers of filing an E.E.O. — upper management no longer considered you loyal, or trustworthy, therefore, you were no longer considered management material. You could never be trusted again. As I was walking back to my office, an overall feeling of weakness engulfed my body. I became dizzy, and was overcome by fear. I went into the doctor's office, and they took my pressure. It was high. I told the nurse what happened, and she suggested that if I did not feel better during the course of the day, perhaps I should go home.

Once I got back to my office, I sat down for awhile. Then, I found myself crying again. I got up, and went into my manager's office, and handed him a slip requesting sick leave for the rest of the day. He looked up and asked me, "Why?" I replied, "I'm not feeling

---

[3] Equal Employment Opportunity

well." He looked at me and said, "What's wrong?" I will never forget the expression on his face, nor the words, "What's wrong?" I did not answer. He signed the slip, and I left for the day. I got into my car, and drove straight to my doctor's office. After explaining to him what happened, he told me that I should take the next couple of days off. He told me that my pressure was up, and apparently, I was under stress. I took his advise. The next day I called my manager, and told him that I would not be in until Monday.

THE PLOT THICKENS

On my return to work the following Monday, my manager asked to see me. When I entered his office, he said to me, "This is the letter I was talking about." He handed me the certified envelope that contained my Letter of Warning. I signed for it, and as I did, I did everything to fight back the tears. He read from his copy. The words were so painful. It read, "You are being charged with *Failure To Perform The Duties Of Your Position. Specifically, Failure to Report An Accident.* This Letter of Warning is being served in an effort to improve your deficiency." I became sick, and angry. Everything hit me at once. I looked straight at him, and thought to myself, "You low life son of a b----." Things were becoming clearer to me. I knew he would do anything to retaliate against me.

As I drove home that day, I felt sick again. I could feel my pressure rising. I tried to get my manager off my mind in order to suppress my anger, but hard as I tried I knew what was being done, was done in an effort to tear me down. Why didn't the supervisor who the injured employee first reported the accident to, and who failed to assist the employee, get a Letter of Warning? Why single me out? I was feeling the first blows of reprisal.

A couple of days later, I received a phone call from an employee who heard what happened to me. He stated that he knew, for a fact, that the same manager who issued me a Letter of Warning, was guilty of the same actions he accused me of — failure to report an accident. The employee stated to me that he had suffered an injury, and when he approached my manager to report the accident, my manager told him to wait until his supervisor came back to report the accident. I asked him to supply me with a statement to that affect. He did. I included the employee's statement in the appeal of my Letter of Warning. I wrote letters to headquarters, including the letter from the employee. Nothing came of it. Later on, I received a letter indicating that my appeal was denied.

THE EVICTION

Several weeks later, I went on vacation. When I returned, my manager asked to see me. I went into his office, and he said to me, "You have been asking for training, and we have decided that you will get it." I felt excited because I thought that I was actually going to receive the training that I had requested. I even managed to smile. My manager looked at me, and said, "Effective one week from today, you are to report to Delivery and Collections." I was shocked. I never asked for such training. I told him that. He said, "It is the training you will get." I told him that I did not want to go. "You have no choice," he said. He was brutal. He wouldn't put it in writing. He practically told me to pack up, and get out. When I asked him for how long, he replied, "Indefinitely."

At that point, I truly understood *violence in the workplace.* I wanted to hurt this man. I wished him every known illness, and some that were yet to be discovered. In short, I wished him dead. Again, I made my visit to E.E.O., filing yet another complaint. Of course, now each trip to E.E.O. became easier. Initially, it was the hardest thing for me to do. But, as time goes by, and as management continues to abuse you, it becomes much easier.

I couldn't sleep; all I thought about was wishing this man harm. I found myself obsessed with hurting him. I was obsessed with finding a way to get even with him. I felt as though *he* got away with murder. I had been forced to leave my position, because he couldn't stand the fact that I had the audacity to question, complain, and cause him to answer to my charges. He took the cowards way out; and upper management condoned his actions. Not one time would he sit down with me, and try to discuss the problem. Instead, it was a cold, silent, good-bye.

I packed my things and left the office.

PURGATORY
"A place, or state, of temporary suffering, or misery"

I reported to Delivery and Collections, as ordered. I felt like a fish out of water. No one was expecting me. I was feeling so depressed, and I didn't want to talk with anyone. I never had any training in the area, and no one was eager to train me. They actually thought of me as being "in the way." Everyone knew I had gotten on someone's sh-- list. Why else would I be there?

One day, a supervisor asked me to go outside with him; he wanted to talk to me. He told me that he could see that I was feeling miserable, and depressed, about something. I broke down. He went on to tell me that, although he was from Delivery, he too, had a story to tell. I gathered from his conversation that everyone who stepped on management's toes were sent to Delivery as a form of punishment. His punishment was a downgrade. He, too, was fighting to get his position back. It surprised me that he would take the time to express concern, and disclose his disappointment with the system, too. Hence, came the nickname, *Purgatory,* for Delivery. I soon began meeting other supervisors who were being punished, and sent to Delivery.

Delivery and Collections is where they send undesirables who are labeled as trouble makers. It used to be that when you stepped on management's toes, you were sent to the Mark-up unit.

Now, everyone was being sent to Delivery. I will never forget his inspirational talk with me. He told me to stop feeling sorry for myself, and fight them with all my might. He also told me not to take it out on anyone else. After ouR talk, I managed to come to work with a different frame of mind, even though I still felt hate, bitterness, and resentment. I managed to get the job done.

We soon parted ways, as I was sent to another station to work.

THERAPY

I began a letter writing campaign. But, no one had the decency to respond. It was as though I never existed. During this time, I began to experience painful feet. My hours were constantly changed. I couldn't sleep at night. My feet began to swell. I was troubled with insomnia, and even though I changed my attitude about my work, thoughts of hurting my manager never left my mind. I would picture him stepping in front of my car, and me pushing down on the gas peddle in an effort to run him over. I knew this was an unhealthy feeling. It had even affected my home life. I became irritable. I gained weight. I was a total mess. I decided to see my family doctor. He referred me to a psychiatrist. I have been in therapy ever since.

After almost a year in Delivery, the manager of the station called me in to tell me that my manager, the one who kicked me out of my position, wanted to see me. I froze with apprehension. I did not know what to expect; I was planning for anything unpleasant.

I first met with my therapist, who advised me how not to lose control. That's what I feared the most.

The day I had to meet with my manager, my car battery went dead. "This is probably an omen," I told myself. However, I got a boost, and headed for the meeting. I had to keep my motor running, so when I went inside to meet with him, I told him that I had to keep my motor running, and asked could we talk outside. He agreed, and as we walked to the parking lot, I began to feel sick.

As we stood by the car, he began to talk. At first, I could barely understand what he was saying; I was feeling so nervous and sick all over. Just seeing him again brought back the pain, and ugliness, of the whole situation. Finally, I heard him say that he was taking me out of Delivery & Collections, and bringing me back. I thought perhaps the letters I was sending all over the place finally paid off. Then he said, "You will be leaving Delivery & Collections, but you will be detailed somewhere else because you are needed there." "I don't want to go anywhere else. The only place I want to go, is back to my position," He said, "You will be receiving a letter from me with instructions on where to report." I told him, "You could have told me that over the telephone." He then went on to tell me how the new person who he placed in my position was doing such a fantastic job.

I never felt such fury. I felt that he deliberately had me meet with him so that he could deliberately berate, and humiliate me. I felt such hate for this man. As he turned away and began to leave, I pictured hundreds of daggers stabbing him in his back. I got into my car, and I don't know how I made it home. I cried all the way.

A couple of days later, I received my instructions to report to my new assignment. When I arrived, I told the manager that I did not ask to come, and that I knew he did not ask for me to come. But I assured him that I would do the best job I could.

My feet had been so damaged from standing, and walking while in Delivery & Collections, that I filed a CA-2.[4] I have had to undergo foot surgery as a result of the damage done to my feet. However, even with supporting medical documentation confirming that my foot condition is job-related, my claim was denied by the Office of Workers Compensation Programs. I have appealed their decision to deny my claim.

---

[4] Notice of Occupational Disease and Claim for Compensation

Since being assigned to this new detail, I have had to file several more E.E.O. complaints against my manager. I have been subjected to ridicule by other postal employees who have commented that I was removed from my position because I was incapable of performing my duties. The embarrassment is endless. Management has kept me from my position for 4 years. There is no mention of my ever being returned. I have been denied opportunities that upper management has allowed others to take advantage of. I have been disgraced, humiliated, embarrassed, and stressed out over this entire experience. I have hired an attorney to represent me at my E.E.O. hearing.

I thank the great staff in EAP,[5] who I visited at the start of my problems. I also thank my personal family physician who confirmed my depression, and acted swiftly in recommending me to a psychiatrist, who along with medication, controlled my anger, and depression. Sessions with him allowed me to laugh once again. I will never forget his words of encouragement. Because of him, I have turned to writing as a release.

As I look back, I am certain that I would not have changed one thing. I will never regret any of my actions. I dared to question, I dared to protest. I made a statement. Although it has caused me great suffering, there is nothing I would have done differently.

Employees who roll over, and play dead, and allow management to continue its abuse, will always be subjected to abuse. It is time to take a stand, and perhaps in numbers, we will make a difference.

---

[5] Employees Assistance Program

CHAPTER X

IS ANYONE LISTENING?

There is a common thread in the interviews you have just read; systemic denial of workers' compensation claims, even though a causal relationship had been established between the work being performed and the injury; medical evidence being blatantly disregarded, outright violation of confidential medical information; threats of loss of jobs with impunity; lack of training for employees; dangerous working conditions; hostile work environments, and ego tripping by management, as they abuse their power of authority. According to the federal worker, to date, no one has found any steadfast answers to ending the types of abusive behavior by management that has been portrayed in *Out In the Open*...

Unfortunately, there are far, far, too many more such stories, and in many cases, the true horrors cannot be told. Take, for example, Sergio. The depth, and full extent of what has happened, and continues to happen to him, simply, *cannot* be told, because postal violence is alive, and thriving, and *no one* is listening. As a result, the abusive behavior that contributes to violence in the workplace continues, and workers like Sergio continue to fear for their lives.

But, is anyone listening?

Postal workers have become weary of Postmaster General Marvin T. Runyon's pronouncement, "We're changing it." According to many federal workers who were interviewed, any element of trust that may have existed, has long since disappeared, simply, because things have not changed, long after the Postmaster General said they would. An electronic technician from the south, spoke about a meeting that was intended to foster efforts between managment, and union, to improve postal working conditions. These efforts were made after accumulated years of ill-treatment, harassment, bullying, and intimidation of the employees in his particular facility. The employee spoke about a parliamentary style of management that exists in the postal service. He said, "This is no way to treat people, no way to run a business."

The Postmaster General has commendably advocated "Zero Tolerance for Workplace Violence." Unfortunately, the violence still exists. It continues to rear its ugly head in many different forms, not just in shooting sprees. The impact, nonetheless, is just as devastating. Take for example, the Florida letter sorter who charged that her supervisor forced her to perform heavy work which eventually led to, among other injuries, a miscarriage.[1] She sued the postal service for $2 million. Then there was the Tennessee mail sorter who became sick after being verbally attacked by his supervisor. The worker subsequently lost consciousness, and the supervisor prevented co-workers from coming to his assistance.[2]

Ironically, according to a news article,[3] "Postal career executives must have 'people skills' — or learn them — to be promoted in the U.S. Postal Service." An Employee Relations Chief states, *"The only crony who will get through is one who is people*

---

[1] *Sun-Sentinel*, September 30, 1993.

[2] *Underwood v. United States Postal Service*, 742 F.Supp. 968 (M.D. Tenn. 1990)

[3] *Federal Times*, June 21, 1993

oriented with a good management track record." In her interview, Caitlin, a former postal supervisor, talked about what constituted a good manager. Simply stated, a good manager is one who abuses craft employees. She states that somehow abusive behavior equates with being a good manager. Another employee who was interviewed, talked about the letter carrier she met when she first went to work for the post office. The letter carrier said she wanted to, "Just get my foot in the door so I can become a supervisor." Well, her dream eventually came true, she became a supervisor. She even dressed differently. Like a supervisor. But it didn't last. She said, "I could not be what they wanted me to be. I could not do the things to craft employees that they wanted me to do. So I asked to go back to carrying mail."

According the Employee Relations Chief, "Executives can be broken down into four categories: people oriented, and successful; people oriented, and not successful; not people oriented, and not successful; and not people oriented, but successful."

A good example of the "people oriented, and not successful" supervisor, was the interview with the postal manager who had advanced up the agency ladder, until he testified against a white postmaster on behalf of a black, female clerk. The clerk charged the postmaster with discrimination, and the manager testified against the postmaster.[4] Even though the manager won his case based on retaliation before the Merit Systems Protection Board, and his back pay was restored, to date that manager continues to suffer retaliation by upper postal management.

As to the, "not people oriented, but successful supervisor," Dr. Wright summed up that analysis this way, "*[t]hey {employees} get involved with certain managers who want to use their power, or abuse their power, in ways that have nothing to do with their supervisory responsibilities. When the employee is drummed out of the agency because, emotionally, they can no longer work, the*

---

[4] *Federal Times*, July 1, 1996

supervisor will find somebody else to hassle somewhere down the line. It's not going to stop."

Postal workers are also drummed out of the agency for other, innocuous reasons, such as "taking baby steps." In a recent television interview,[5] after approximately 18 years with the postal service, a letter carrier spoke about her removal from the agency because managment felt she did not take large enough steps while delivering the mail. Her husband photographed management following his wife as she delivered the mail. In her interview she said, "Management was walking slower than me."

Out In The Open... addresses any and all workers who have ever suffered abuse from management. However, according to the Neurologist, Psychologist, and Social Worker who were interviewed, it appears that management abuse of injured workers is particularly endemic to the federal sector, and more specifically, the postal service.

In an effort to address their growing concerns, in 1992, a group of injured federal workers gathered, and joined the National Coalition of Injured Workers. Their efforts to effectuate change about issues which affected them, included lobbying Congress, and agencies, to enact *meaningful* legislation designed to re-vamp U.S. Workers Compensation laws that have become outdated, and useless. Again, no one is listening. At one time there was a Post Office and Civil Service Committee. Unfortunately, even the valiant, sincere and well-intended efforts of the Committee's former chairman were not enough. The power of the postal service was *too* powerful, and the straw committee fell.

In his interview, Sergio summed it up best when he said, *"[t]his isn't an honest agency. It's not a fair agency. When you get an agency this powerful, this strong, something needs to be done; I know can't nobody stop what's going to happen to me. I know that the system, the postal service, is too powerful..."*

---

[5] <u>Good Morning America</u> *ABC*, September 16, 1997

SUGGESTED SOLUTIONS

None of the suggestions listed below are legal advice and should not be taken as legal advice, but rather, common sense approaches to some problems that may arise.

Federal injured workers ask, "If the Federal Employees Compensation Act,[6] and the Office of Workers Compensation Programs,[7] is not, will not, and cannot help, who then, will help? The answer is, *You must help yourself.* Help yourself by first, informing yourself as to what rights you have; identify and refuse to accept sexual harassment, and verbal abuse; recognize signs that lead to emotional stress; learn how to seek help; know what to do, where to go, and the type of doctor to see when you suffer a physical or an emotional injury. Most of all, ask yourself, "Am I willing to allow the hazards of this job to consume my life, and the lives of my family?"

1. What Are Your Rights?

You have the right to work in a violent-free, hostile-free, and discriminatory-free work environment. No one contracts to go into a workplace to be abused either by co-workers or management. Three of the best pieces of reading material you can ever have in your library are: *A Guide to Federal Sector Equal Employment Law and Practice, by Ernest C. Hadley* (Dewey Publications, Inc), *A Guide to Merit Systems Protection Board Law*

---

[6] FECA
[7] OWCP

and Practice, by Peter Broida (Dewey Publications), and *The Federal Employee Procedure Manual, Part 2 - Claims* (U.S. Department of Labor, Office of Workers' Compensation Programs, Washington, DC). These three books contain a wealth of information. With this material, you will have the same information as the agencies, and you will know the rules of their playing field. Buy them, and read them *before* you need them; and if you are already engaged in dispute, read them to assist yourself, and your representative. There are no guarantees, but at least you will give yourself a heads up, and the opportunity to *broaden* your chances of securing a favorable claim.

2. **Refuse to Accept Sexual Harassment And Verbal Abuse**

Sexual harassment is more common than you may think. In a report to the President, the Merit Systems Protection Board conducted a survey, and found that in 1994 the government average of women who received unwanted sexual attention was 44 per cent, and for men, 19 per cent.[8] While this report was informative, it is interesting to note that, in the section concerning "How Much Harassment Is Occurring," among the federal agencies named, the United States Postal Service, is conspicuously absent.[9] *A Guide to Federal Sector EEO Law & Practice* (1996) has a comprehensive section on Sexual Harassment including, what elements constitute a *prima facie* case, the "reasonable woman" standard, consensual conduct, prevention, and cessation of conduct.

Many things constitute sexual harassment, including actual or attempted rape, or sexual assault; unwanted letters, telephone calls, or materials of a sexual nature; unwanted sexual teasing, jokes, remarks or questions; unwanted deliberate touching, leaning over, cornering, or pinching. In The Postal Supervisor, readers

---

[8] *Sexual Harassment in the Federal Workplace: Trends, Progress, Continuing Challenges*, October, 1995, page 15
[9] Id, page 15

were informed that an "employee 'touching' memo" had been issued by the New York area.[10] According to the PS, the memo put supervisors on notice that the courts, "will not defend an EAS supervisor acting within the scope of his or her employment, if the facts clearly reveal that 'inappropriate touching' of another employee, especially a subordinate, occurred, even if it was originally perceived to be consensual."

In spite of the memo, employees ask, "Why does the matter have to reach to the courts? Why aren't such matters dealt with efficiently and resolved at the administrative levels, particularly when the abusive behavior is clear?" These are the questions asked by Caitlin, Cynthia, Valerie and Arnie. All of them were sexually abused, in one form or another, and nothing was done at the administrative level in either case. Today, neither of them works for the postal service, because they have been permanently damaged by their experiences as employees with the U.S. Postal Service.

In some cases, the employee prevails — as it relates to the issues of their case. But emotionally, they, and their family members, may never be the same. Take for example Susan, who worked as a mail handler in a large post office in the south. Her story was portrayed in the article, *Friendly Faces That Can't Smile.*[11] The article focuses on abuse of power of a manager who continually barraged Susan with unwanted sexual advances. In Susan's situation, as in most situations concerning abusive behavior by management, she was stonewalled — how do you complain to one manager about another manger's sexual harassment when they are fishing, golfing and bowling buddies?

Sexual harassment is yet another form of violence, and its impact is devastating, not only to the victim, but to co-workers and

---

[10] March 21, 1997, Volume 88, Number 6
[11] *Postal Life*, September/October 1996

families as well. Employees sometimes make their feelings know, but, [12] again, no one is listening.

3. Recognize Signs That Lead To Emotional Stress

According to the physicians interviewed, signs of stress include, lack of sleep, irritability, crying, eating disorders, and depression. Second to sexual harassment, federal employees most frequently cite retaliation by management as their source of emotional stress. Two most frequently cited reasons for retaliation by management are: employees' engaging in federally protected activity, such as, filing an EEO complaint, or a grievance, or being a witness in an EEO proceeding, and filing a workers' compensation claim. Remember Valerie, who suffered retaliation, and was eventually downgraded to cleaning toilets when she filed an EEO complaint against the supervisor who sexually harassed her?

When considering filing a complaint based on retaliation, the best way to establish a *pattern* of retaliatory behavior is to keep a journal, and on a daily basis, document log what you consider to be patterns of discriminatory behavior. Purchase a black and white composition notebook. On the front cover, put your name, and date of the first entry. Write on both sides of the pages. Date each page at the top, number each page at the bottom. Include names, dates and places of incidents. If possible, immediately get witness statements. And never, tear any pages from the note book. Continually provide yourself with a pattern of information. If you engage in a federally protected activity that your management does not like, and things begin to happen so close in time so as to give rise to an inference of retaliation, your log of events can turn out to be your best friend.

---

[12] *Sexual Harassment in the Federal Workplace: Trends, Progress, Continuing Challenges* October, 1995, Page 23

4. Seek Help

Take advantage of the Employees Assistance Program.[13] The first 12 visits are free. This program was initially set up to assist employees who had drug, behavioral, or work-related problems. According to Gabriele, in the past, the postal service has gone out of it's way, and has done an outstanding job, in gaining employees' mistrust, and decreasing their moral to a deplorable state. In the past years, in a last ditch effort to restore employees faith in the organization, the postal service began sending employees yearly employee opinion surveys to their home. When that did not seem to work out, for whatever reasons, perhaps not enough participation, they then began to hand out the employee surveys on the clock. They even gave the employee time to fill them out — under the watchful eye of an employee from the personnel department. Gabriele says, "However, this method appeared strange to the employee, since the forms were not to be signed, and no one was to know the identity of each participant. Clearly, if the employee made any negative comments, or derogatory statements, the employee has set him/herself up for reprisal. After all, the form contained such questions as, the department you work in, your level, years of service, race, etc., just a few questions that can "clearly" identify the employee.

According to one employee, "Heck, an employee is sent to oversee this whole fiasco. Of course, you did have the option not to fill out the questionnaire, but guess what happens? You have now been identified by the overseer as not being cooperative."

Then the EAP was re-vamped. Instead of the counselors being postal employees, now the whole operation has been contracted out to a private firm. The counselors that represent the EAP are individuals who have a Masters in Social Work. They

---

[13] EAP

cannot prescribe any medications, because they are not doctors. They are there to listen to you, and to advise you.

Gabriele used the EAP under the re-vamped system. She said, "It becomes an emotional experience, because you are encouraged to speak up and let your feelings flow. Like any psychotherapy session." Psychotherapy is treatment for problems that are assumed to be of an emotional nature. Therapy is administered by a trained person who deliberately establishes a professional relationship with a patient for the purpose or removing, modifying or retarding existing symptoms, or reversing patterns of behavior, and of promoting positive personality growth and development.[14]

Remember Arnie. He did not want to seek therapy because of the stigma attached. And Sergio, he did not want to go to therapy because of his mistrust of the system. But in both cases, they felt better by talking about their problems, and relieving the pressures of their frustration. It gave them an opportunity to clear their heads so that they can move on with their lives.

The purpose of the EAP is to counsel employees in the areas of drug counseling, legal referral, stress counseling, financial counseling and job-related counseling. They will also assist family members. And, you are permitted 12 free visits. The first visit is on the clock. All other visits are on your own time.

In her interview, Gabriele talks about her experience with EAP. "Unfortunately, the perception of EAP by postal employees is that EAP is a referral service just for drug users. (*Remember, Arnie said the same thing*). Employees distrust management, and although they are told that their visits are confidential, they do not believe this to be true. I attended 12 visits to EAP last year, and during my visits, although my counselor never mentioned names, she made it known to me that she had heard so many horror stories from postal employees. Although the EAP handle other agency's

---

[14] Definition from *Psychiatric Dictionary*, 5th Edition, Oxford Press, 1981

employees, not only have they seen more postal employees, but the horror stories that postal employees have related to them, were the worse. My counselor claimed that not only did she counsel craft employees, but there were just as many EAS (management staff) employees. The cases she handled were not only of severe depression and stress, but also of potentially violent situations.

While the EAP staff are sworn to confidentiality, the only time they break their code of silence is in the event that an employee indicates that they are a threat to another individual, and that they are contemplating violence. EAP will then notify the person who may be in danger. EAP staff does not take notes, nor is there a tape recorder present.

Stress, and depression, are the major problems that exists in the postal service. It is so depressing that it caused my counselor to resign from the EAP services. She now counsels individuals from her home. She also prefers not to take on postal employees, because, as she has stated to me, it is too depressing for *her*. She always found herself in a "no-win situation." She confessed to me that upper management is absolutely out of control. They have thrown the rule book out of the window, and have caused moral to deteriorate to such a degree that it seems to be un-repairable. Individuals that display severe cases of emotional stress, or depression, are referred to a psychiatrist.

The EAP counselors are sincere in their effort to help, but they confess that the postal service is on the verge of disaster. A time bomb waiting to go off. They are never surprised when they hear of violence in the workplace at the post office, such as the recent killings that have taken place. Other agencies which EAP represents are working with their employees, and have put together programs in order to regain their employee's confidence, and trust. They work hard at overcoming problems that caused their employees to seek help in the first place. However, because the postal service is run like a military facility, and lacks creativity and innovation, it is practically impossible for supervisors to show any human compassion. *(Remember what Dr. Tai said about the military mentality)* The postal service punishes the supervisors who do not punish employees. It is a Gestapo, in the military manner in which it its run, even to the point of using military time. This places any employee who is unfamiliar with the military in an uncomfortable, and strange environment the minute they walk through the door.

It is best that an employee refer themselves to the EAP for assistance without notifying their supervisor. Since the first visit is on the clock, the supervisor is aware of it. Now the employee has labeled themselves as troubled. No matter what management says, the employee is labeled. Everyone that the supervisor tells, will know. Everyone will draw their own conclusion, as to why the employee needed assistance. It is indeed, a sad state of affairs. By not notifying the employer, EAP does not communicate your visits to them. It can be a good experience for the employee, especially if they cannot afford private counseling. EAP gives them an opportunity to let off steam, and possibly, when they have finished their counseling, leave with a feeling that someone cared, that someone understood what they have been going through, and possibly, gain a friend like I did. It can be a rewarding experience."

5. **What To Do When You Are Injured On The Job**

Immediately report all injuries. Section 821.311 of the Employee Labor Relations Manual (E.L.M.) addresses the issue of Reporting An Accident. The general requirement of the E.L.M., states that the manager or supervisor of the employee, or operation involved must, (a) report all accidents on Form 1769, Accident Report, within 24 hours of the date of the accident, the diagnosis of illness, or the date the manager/supervisor was notified of the situation, and (b) provide a copy of Form 1769 to the employee in the accident upon written request. Completion of the form is mandatory, regardless of tort claim action, or the requirements of the Federal Employees Compensation Act. What the E.L.M. doesn't state is that the person who was involved in the accident, whether at fault or not, more than likely, will be issued a "Letter of

Warning." A postal supervisor who was interviewed said, "All letter of warnings are written to read, 'This Letter of Warning is being issued to you in an attempt to correct your deficiency. Any further deficiencies of a similar nature will result in more severe disciplinary actions, including suspension, or removal from the Postal Service.' What a frightening thought! Where is the deficiency? Suppose you trip and fall, or are hit by a piece of moving equipment, where were you deficient?"

Mail Carriers are given letter of warnings for having either at fault, or no fault accidents. Sometimes they are placed on suspension; at worse, they are served with letters of removal, depending on the severity of the accident. It doesn't matter who was at fault, there will surely be some form of discipline issued. *(Remember Mikie whose finger was severed while he slotted mail).*

Many new employees who have accidents during their 90 day evaluation period, are fired. Those employees that do not fear upper management, file a grievance, or an E.E.O. complaint, in an attempt to regain their positions. However, more than likely, they will not prevail. Therefore, many new employees who do suffer injuries on the job, do not report the injury, for fear that they will be terminated. *(Remember Valerie)* The supervisor said she had two employees who were being sexually harassed by a Mail Carrier who, not only thought of himself as a ladies man, but had a record of sexual harassment. The two employees became traumatized by the experience. The employees went as far as filing an E.E.O. complaint, only to drop it because they were told by upper management that if they pursued the complaint, they would be terminated; which meant they would never be hired by the Postal Service again. This left them fearful, however, management did nothing to relocate them away from the accused employee, (or yet, fire the accused employee). Instead, the employees were left to be continuously laughed at, and taunted by other Mail Carriers. This situation caused them great stress, resulting in time being taken off from work due to a job-related stress condition. Because they were new employees, they did not want to file a CA-2[15] for fear of losing their jobs. Consequently, they were labeled "trouble makers" *(Remember Dr. Wright)*, and when their 90 day casual assignment was over, they were not kept on the rolls. Instead, they were told that because of their attendance, they were no longer needed. Now these individuals not only suffered the loss of a job, but also an ongoing stress related condition that rendered them helpless.

The situation is no different for permanent, regular employees. Many employees fear reporting any injuries because they fear reprisal by management. On June 5, 1981, the acting regional postmaster general of the Northeast Regional Office of the USPS issued a memo in which it addressed the subject: Discipline for Safety Rule Violations. The memo was issued to postmasters, district managers, general managers, and managers. In part, the memo states, "The Federal Employee Compensation Act, and our own policies <u>prohibit</u> taking action that <u>discourages</u> reporting claims for compensable injury with the Office of Workmen's Compensation. However, in sixteen years, nothing has changed. Today, employees continue to suffer reprisal for filing workers' compensation claims.

According to the injured employees interviewed, the injury compensation office is adept at conveniently, losing, mis-filing, and in many instances, discarding workers compensation claims. *(Remember Cynthia)* Regulations impose penalties upon *"Any person charged with the responsibility of making reports in connection with an injury who willfully fails, neglects, or refuses to do so, induces, compels, or directs an injured employee to forego filing a claim, or willfully retains any notice, report, or paper required in connection with an injury, is subject to a fine or not more than $500, or imprisonment for not more than one year, or both."*[16] This regulation continues to be violated. According to one

---

[15] Notice of Occupational Disease and Claim for Compensation
[16] 20 *C.F.R. 10.23(c)*

employee, "I always CYA {cover my a—}. I always send a copy of my Ca-1, Ca-2, Ca-2a, CA-8, and whatever other forms that I complete, directly to OWCP. Of course, I keep copies for myself. I log it in my diary as to when I sent it, and how I sent it. I write a cover letter to OWCP letting them know the reason I am sending them a copy of my claim is because of management's hostile attitude, and the track record of the injury comp unit losing claims. I feel this is the only way OWCP will have a record that I did file a claim for a job-related injury."

6. Finding Medical Treatment

According to Dr. Johns, a neurologist who was interviewed, finding medical treatment, and the right medical specialist, can be accomplished in several ways. She stated that while the yellow pages, and lawyers can be a source of referrals, the most common referrals are done by word of mouth, and through the HMO primary physician. Dr. Johns cautions, however, that when seeking a workers' compensation doctor, ask *many* questions, such as the physician's experience with, not only workers' compensation claims, but specifically, *federal* workers' compensation claims. Look for signs as to whether the doctor will take you, and your injury, serious, Take note: is the doctor is receptive? Dr. Johns says, "The injured employee should worry if, first, they have a 'gut reaction' that the doctor has an aloof attitude, and second, if the doctor minimizes the symptoms."

Dr. Johns further stated that it is important to find the right doctor for the right injury. For example, a neurologist treats chronic pain, such as head, neck, back and nerve damage. Nerve damage may manifest itself in numbness, tingling, a pins and needles feeling, and any burning or electric shock type feelings. She goes on to state that, depression is secondary to chronic pain, "Anyone who suffers with pain for longer than six months, invariably become depressed." Some symptoms of depression include, staying up all night and irritability. In these situations, it is harder for the patient to heal. So now, the employee is not only being treated with pain medication, but anti-depressants as well.

"There seems to be a lack of sympathy and compassion among the hierarchy in the post office," Dr. Johns states. "It does not appear to be a nurturing atmosphere; promotion appears to be based on seniority, rather than skill." She further states, "There seems to be a need for a less stressful environment; more attention should be given to the employees feelings; the employer needs to factor in the value of human life. There needs to be more support groups, psychologists, and open meetings for the employees to vent their emotions."

She also noted that postal employees appear "not to be trained to find other jobs; they are not in competitive fields; theirs appear to be dead-end jobs." In the treatment of her postal patients, Dr. Johns concludes that it has been her observation that, because of the egregious wrongs done to injured postal employees, at some point, the employee wants revenge to right the wrongs done to them. She states, "Perhaps they stay *(at the post office)* because of the pension plans, I don't know. In the private industry, they just quit, and find other jobs."

The psychiatrist and psychologist usually work together as a team. While the psychiatrist dispenses medication, the psychologist is the person with whom the individual meets with regularly to discuss their problems in-depth; they move toward working out solutions, and they learn stress management. *(Remember Dr. Tai).* Be aware, however, that when selecting a psychiatrist, OWCP, and the Employees Compensation Appeals Board (ECAB), *only* recognize those who are Board Certified. It is extremely helpful to find a psychiatrist-psychologist team, whose reports will be supportive of each other, since, OWCP or ECAB don not give probative weight to the reports of a psychologist standing alone.

According to an article in *The Postal Record,* "More than any other reason, claims for OWCP benefits fail because of lack of rationalized medical evidence..."[17] According to the FECA manual, causal relationship is defined as a direct cause of the injury to the factors of your employment.

So, what are *factors* of your employment? You are a mail carrier, and a factor of your employment is to slot mail. While slotting mail, your finger is severed. *That* is a direct cause of the injury as a result of the factors of your employment. But what about the not so obvious causal relationships, such as carpal tunnel relative to the repetitive wrist motion from keying mail. In 1994, ergonomic standard regulations were introduced by the Labor Department's Occupational Safety and Health Administration.[18] The standard is designed to affect any worker whose job could cause musculoskeletal disorders such as back strain, and carpal tunnel syndrome. *(Remember Cynthia)*

If you sustain a back related injury, under the FECA, the services of chiropractors may be reimbursed only for treatment consisting of manual manipulation of spine to correct subluxations as demonstrated by X-ray to exist. But does the "rationalized" report from your treating physician win your claim hands down? Not necessarily. More than likely, regardless of what your treating physicians writes, you, your doctor and your representative will jump through many hoops. If your doctor cares about your well-being, although he/she may become frustrated with the endless bouts back and forth with the agency and OWCP, he/she will hang in their with you until the end. If your representative is worth his/her salt, they too will fight to the finish until every element of your due process in the administrative forum has been exhausted. Therefore, you too, must keep your self-esteem high, your pride unwavering, and hang in there for the long haul. The price of victory is high.

Language in the physicians report is everything. If their language does not reflect the language contained within the FECA

---

[17] *The Postal Record*, Compensation Department, May 1996
[18] *Federal Times*, November 21, 1994

manual, ergo FECA's interpretation of an injury, not the physicians, expect your claim to be automatically controverted by the agency, and denied by OWCP. Remember, you always have the right to appeal before ECAB. However, prior to appealing to that forum, make sure you have all the supportive medical documentation possible, since ECAB does not review *any* new evidence. They are sharp, they will know if you try to introduce something new. So don't.

Perhaps, the most difficult claim to prove is the occupational disease or illness for emotional stress. These claims seem to always be controverted by the agency, and denied by OWCP. The case of <u>Lillian Cutler</u> 28 ECAB 125[19] is *always* used by OWCP to deny a claim for emotional stress. The history of the events that led to the emotional stress as taken by the psychiatrist, and psychologist working in conjunction with the psychiatrist, is critical. Just as critical is the "language" used by the psychiatrist. It must be FECA focused language.

Finally, according to ECAB, "An injury does not have to be confirmed by eyewitnesses in order to establish the fact that an employee sustained an injury in the performance of duty, as alleged, *but*, the employee's statements must be consistent with surrounding facts and circumstances, and his/her subsequent course of action."[20]

Unfortunately, there are no guarantees. Winning your case will not be a stroll in the park, depending upon the nature of the injury. Therefore, be informed, *before* an injury occurs.

---

[19] Another excellent resource is the Digest and Decisions of the Employees' Compensation Appeals Board. Go to the Government Publications section of your local library. Currently, there are 44 Volumes.

[20] <u>Constance G. Patterson</u>, 41 ECAB 206 (1989); <u>Bill H. Harris</u>, 41 ECAB 216 (1989).

In the event you do have an accident:

1. Report it immediately. If a supervisor, or manager does nothing, ask to see a shop steward.
2. Make sure that you immediately take any witnesses names. It *helps* to substantiate your claim.
3. Document everything.
4. Let the supervisor, or manager take you to the doctor.
5. After your initial visit to the agency doctor, *immediately* make an appointment to see a doctor of *your* choice, not the agency's choice.
6. Take control of your situation. Do not wait for Injury Compensation to make appointments for you. They will only send you to a doctor that works for them.
7. *Do not* discuss your injury with anybody, except your doctor. Your co-workers can sometimes be your worse enemies.
8. Be sure to get statements from any witnesses as soon as humanly possible. Statements dated long after the incident are often viewed as suspicious.
9. Respond to all correspondence from OWCP in a *timely* manner. If you are physically unable to do this yourself, get a family member, or a friend, to help you. You will be working within certain time restrictions. *Make copies of everything before you submit it.*
10. Keep all doctor appointments.
11. Do not let Injury Compensation intimidate you.

An important note: you will be asked by OWCP to write a statement of what occurred when the accident took place. Take care that you are accurate in detail, as you must be consistent as to the details when you are questioned at a later time, perhaps at a hearing.

7. How To Find A Professional Representative

Seek professional legal representation if necessary. Be aware that *many* lawyers will not take federal workers' compensation cases, or E.E.O. cases involving the postal service. A number of persons interviewed, stated that they contacted as many as 22 lawyers, who simply would not 'touch' a post office case. Many injured employees have become frustrated, and have dropped their claims because of this. This is what the agency hopes for. Don't be discouraged, there are attorney's, and claimant representatives who will take federal workers' compensation, and E.E.O. cases.

When you find someone, a*sk questions*. You want a representative who is well versed in not only federal law, but federal workers compensation procedures, FECA, OWCP, and E.E.O. complaints for *federal employees*. You also want a representative who knows, and understands the procedures for the Merit Systems Protection Board, particularly since this administrative forum deals with veterans; and the National Labor Relations Board. In difficult cases, you may come in contact with all of these agencies. It is not uncommon for some cases to go the gamut, including, retirement disability, unemployment and social security disability. Ask your representative how knowledgeable he or she is in these areas. Don't be afraid to ask them for references from *federal* clients whom they have assisted in any of these areas, or the area(s) similar to your case(s). You want to know if they have a track record for representing federal injured workers, since the administrative procedures that they operate under are very different from state laws and regulations.

Also, because postal employees fall under federal guidelines, it is not necessary that you and your representative live in the same state.

In addition to word of mouth, an excellent resource for identifying a representative is the *National Association of Federal Injured Workers.*[21] They also have a monthly newsletter, *The Band Aid*, which keeps you abreast of what's going on with legislation concerning federal injured workers. There is also a section called Postal Bull. The Band Aid also offers very useful manuals and tapes.

8. Use Your Union

As a dues paying member of your craft, put your union's feet to the fire. Demand fair representation. Fair representation is the duty of the union to serve the interests of all members, without hostility, or discrimination toward any, and to exercise its discretion with complete good faith, and honesty, and to avoid arbitrary conduct.

Every craft has an Agreement with the U.S. Postal Service. There is also a Local Memorandum of Understanding. These Agreements are invaluable to every federal employee. Read them. Learn them. If you don't understand Article 15 of the NALC Grievance-Arbitration Procedures, get your shop steward to explain it to you. If you don't understand Article 16 of the NPMU Discipline Procedures, get your shop steward to explain it to you. *Know* the role, and responsibilities, of your shop steward. Know who your union President, and union Vice President is, and their roles and responsibilities. If you don't know who they are, or what they are supposed to be doing for you, you can't put their feet to fire.

When actions by your union result in a discipline for you such as a 14-day suspension or removal — contact your local National Labor Relations Board and ask for a form that allows you to lodge a charge against your labor organization, or its agents, under Section 8(b) of the National Labor Relations Act.

---

[21] (541) 472-8940

CONCLUSION

Many postal workers, have retired on disability, or are on the workers compensation rolls.[22] With the exception of two former postal workers interviewed for *Out In the Open...*[23] all are on either, workers compensation and social security disability, or retirement disability and social security disability.

On June 19, 1996, many branches of the National Association of Letter Carriers joined picket lines to protest, among other issues, termination of discussions about privatizing some parts of the agency.[24] Many postal workers, non-postal workers, and doctors who were interviewed for *Out In The Open...* believe that the only way change will take place for injured workers, is through privatization. As one injured worker put it, "Privatization may mean that a lot of postal workers will lose their jobs. But in my opinion, the workers most likely to loose their jobs are those that don't do anything, anyway. The workers that will keep their jobs are those who are experienced, hard-working, take pride in their work, are interested in serving the customers, and who will make a profit for the company."

The general consensus among postal workers is that, whatever decisions are made to effectuate change, one thing is clear, business cannot continue as usual. There must be a change, by any common sense means necessary.

---

[22] *Federal Times*, September 16, 1996, p. 3
[23] 15 federal workers were interviewed for *Out In The Open...*
[24] *Federal Times*, July 1, 1996, p. 11

BIBLIOGRAPHY

A GUIDE TO MERIT SYSTEMS PROTECTION BOARD LAW AND PRACTICE, by Peter Broida

A comprehensive guide about the MSPB forum. Dewey Publications, Inc. 1996

A GUIDE TO FEDERAL SECTOR EQUAL EMPLOYMENT LAW AND PRACTICE, by Ernest C. Hadley

A comprehensive guide about the EEO forum. Dewey Publications, Inc. 1996

BAND AID, THE, by The National Association of Federal Injured Workers

An informative monthly newsletter with its focus on moving Congress to pass bills to reform FECA ,and to restore the rights to injured federal worker.

FEDERAL EMPLOYEES PROCEDURE MANUAL, PART 2 - CLAIMS

A Manual published by the U.S. Department of Labor, Office of Workers' Compensation Programs, Washington, DC

FEDRAL TIME, THE

An informative newspaper with a focus on federal workers, and federal concerns, published every Monday by Army Times Publishing Co.

NATIONAL LABOR RELATIONS ACT

A 40 page booklet which contains both the National Labor Relations Act, and the Labor Management Relations Act.

POSTAL LIFE

A magazine published by Corporate Relations, Washington DC 20260-3100, for the nation's postal employees.

POSTAL SUPERVISOR, THE

A weekly membership publication of the National Association of Postal Supervisors (NAPS)

ABOUT THE AUTHOR

Angela V. Greene, J.D. is a native New Yorker having earned her Bachelor of Arts Degree in Communications from Antioch University in Yellow Springs, Ohio and her Juris Doctor Degree from the City University of New York Law School at Queens College in Flushing, New York.

Prior to law school, Angela was a radio producer with National Public Radio in Washington, D.C. where she produced eight documentaries, now housed in the Schomburg Center for Research and Black Culture in New York City. She was nominated for the Ohio State Award for her sound portrait of Tom Feelings, the artist; she was the first Associate Director for the Journalism Institute at Howard University, was Press Secretary for Senator Lilliana Belardo in St. Croix, US Virgin Islands, and was instrumental for the Senator capturing the highest number of votes when she ran for her second term in office.

After law school, she worked with the Legal Aid Bureau in Maryland, the Prisoners' Legal Services and Legal Aid Society in New York, and was Director of the National AIDS Minority Program for the National Urban League in New York. Angela was also a consultant for the National Congress of Neighborhood Women in New York and was the Public Relations Director for Piazzaz in Baltimore, MD. As an invited guest speaker for various organizations, she has traveled to New Orleans, South Carolina, California, New York, Washington, D.C., Virginia, Florida, Minnesota, and Wisconsin and has addressed audiences at Rutgers University, Howard University and the University of New Rochelle, as well as several high schools in New York and Philadelphia. She volunteers for children's hospitals and Junior Achievement.

In 1992 Angela began successfully representing federal injured workers in their various claims before the Office of Workers' Compensation Programs, the Employee's Compensation Appeals Board, the Equal Employment Opportunity Commission, the Office of Personnel Management, the U.S. Merit Systems Protection Board, the National Labor Relations Board, the Social Security Administration and in unemployment claims. In 1995 she was instrumental in changing Office of Workers' Compensation Programs law in the case of <u>Cook v. U.S. Postal Service</u> (ECAB 1995).

Angela lives in Philadelphia, PA, is the mother of two sons, one daughter, four grandchildren, and three great grandchildren. She enjoys traveling, reading, archery competition, long walks on the beach, and children.

NOTES

NOTES